SALINE AGRICULTURE
Salt-Tolerant Plants for Developing Countries

Report of a Panel of the
Board on Science and Technology
for International Development
Office of International Affairs
National Research Council

National Academy Press
Washington, DC 1990

NOTICE: The project that is the subject of this report was approved by the Governing Board of the National Research Council, whose members are drawn from the councils of the National Academy of Sciences, the National Academy of Engineering, and the Institute of Medicine. The members of the committee responsible for the report were chosen for their special competences and with regard for appropriate balance.

This report has been reviewed by a group other than the authors according to procedures approved by a Report Review Committee consisting of members of the National Academy of Sciences, the National Academy of Engineering, and the Institute of Medicine.

The National Academy of Sciences is a private, nonprofit, self-perpetuating society of distinguished scholars engaged in scientific and engineering research, dedicated to the furtherance of science and technology and to their use for the general welfare. Upon the authority of the charter granted to it by the Congress in 1863, the Academy has a mandate that requires it to advise the federal government on scientific and technical matters. Dr. Frank Press is president of the National Academy of Sciences.

The National Academy of Engineering was established in 1964, under the charter of the National Academy of Sciences, as a parallel organization of outstanding engineers. It is autonomous in its administration and in the selection of its members, sharing with the National Academy of Sciences the responsibility for advising the federal government. The National Academy of Engineering also sponsors engineering programs aimed at meeting national needs, encourages education and research, and recognizes the superior achievements of engineers. Dr. Robert M. White is president of the National Academy of Engineering.

The Institute of Medicine was established in 1970 by the National Academy of Sciences to secure the services of eminent members of appropriate professions in the examination of policy matters pertaining to the health of the public. The Institute acts under the responsibility given to the National Academy of Sciences by its congressional charter to be an adviser to the federal government and, upon its own initiative, to identify issues of medical care, research, and education. Dr. Samuel O. Thier is president of the Institute of Medicine.

The National Research Council was established by the National Academy of Sciences in 1916 to associate the broad community of science and technology with the Academy's purposes of furthering knowledge and of advising the federal government. The Council operates in accordance with general policies determined by the Academy under the authority of its congressional charter of 1863, which establishes the Academy as a private, nonprofit, self-governing membership corporation. The Council has become the principal operating agency of both the National Academy of Sciences and the National Academy of Engineering in the conduct of their services to the government, the public, and the scientific and engineering communities. It is administered jointly by both Academies and the Institute of Medicine. The National Academy of Engineering and the Institute of Medicine were established in 1964 and 1970, respectively, under the charter of the National Academy of Sciences.

The Board on Science and Technology for International Development (BOSTID) of the Office of International Affairs addresses a range of issues arising from the ways in which science and technology in developing countries can stimulate and complement the complex processes of social and economic development. It oversees a broad program of bilateral workshops with scientific organizations in developing countries and publishes special studies of technical processes and biological resources of potential importance to developing countries.

This report has been prepared by a panel of the Board on Science and Technology for International Development, Office of International Affairs, National Research Council. Staff support was funded by the Office of the Science Advisor, Agency for International Development, under Grant No. DAN 5538-G-SS-1023-00.

Library of Congress Catalog Card No. 89-64265
ISBN 0-309-04189-9

S088
Printed in the United States of America

Preface

Populations in developing countries are growing so quickly that the land and water are unable to sustain them. In most developing countries, prime farmland and fresh water are already fully utilized. Although irrigation can be employed to bring land in arid areas into production, it often leads to salinization. In some countries, the amount of newly irrigated land is equalled by salinized irrigated land going out of production. Moreover, irrigation water is often drawn from river basins or aquifers shared by several countries, and friction over its use is common.

Salt-tolerant plants, therefore, may provide a sensible alternative for many developing countries. In some cases, salinized farmland can be used without costly remedial measures, and successful rehabilitation of degraded land is usually preferable, in terms of resource conservation, to opening new land. Groundwater too saline for irrigating conventional crops can be used to grow salt-tolerant plants. Even the thousands of kilometers of coastal deserts in developing countries may serve as new agricultural land, with the use of seawater for irrigation of salt-tolerant plants. These plants can be grown using land and water unsuitable for conventional crops and can provide food, fuel, fodder, fiber, resins, essential oils, and pharmaceutical feedstocks.

This report will cover some of the experiences and opportunities in the agricultural use of saline land and water. The purpose of this report is to create greater awareness of salt-tolerant plants—their current and potential uses, and the special needs they may fill in developing countries—on the part of developing country scientists, planners, and administrators, and their counterparts in technical assistance agencies.

Introducing new crops is always risky. Each species has its own peculiarities of germination, growth, harvest, and processing. When unfamiliar plants are launched where land, water, and climate are hostile, difficulties are compounded. Salt-tolerant plants will require special care to help meet the needs of developing countries, but, given their promise, this attention seems increasingly justifiable.

Preparation of this report was coordinated by the Board on Science and Technology for International Development in response to a request from the U. S. Agency for International Development. I would like to acknowledge the contributions of the Panel, the many scientists who reviewed and revised the manuscript, and, in particular, to thank James Aronson and Clive Malcolm for their generous assistance.

Griffin Shay
Staff Study Director

PANEL ON SALINE AGRICULTURE IN DEVELOPING COUNTRIES

J. R. GOODIN, Texas Tech University, Lubbock, Texas, *Chairman*.
EMANUEL EPSTEIN, University of California, Davis, California, USA
CYRUS M. MCKELL, Weber State College, Ogden, Utah, USA
JAMES W. O'LEARY, Environmental Research Laboratory, Tucson, Arizona, USA

Special Contributors

RAFIQ AHMAD, University of Karachi, Karachi, Pakistan
JAMES ARONSON, Ben Gurion University, Beer-Sheva, Israel
AKISSA BAHRI, Centre de Recherches du Genie Rural, Ariana, Tunisia
ROLF CARLSSON, University of Lund, Lund, Sweden
JOHN L. GALLAGHER, University of Delaware, Lewes, Delaware, USA
H. N. LE HOUEROU, CEPE/Louis Emberger, Montpellier, France
E. R. R. IYENGAR, Central Salt and Marine Chemicals Research Institute, Bhavnagar, India
C. V. MALCOLM, Western Australia Department of Agriculture, South Perth, Australia
K. A. MALIK, Nuclear Institute for Agriculture and Biology, Faisalabad, Pakistan
J. F. MORTON, Morton Collectanea, Coral Gables, Florida, USA
DAVID N. SEN, University of Jodhpur, Jodhpur, India
N. YENSEN, NyPa, Inc., Tucson, Arizona, USA
M. A. ZAHRAN, Mansoura University, Mansoura, Egypt

National Research Council Staff

GRIFFIN SHAY, *Senior Program Officer, Staff Study Director*
NOEL VIETMEYER, *Senior Program Officer*
F. R. RUSKIN, *Editor*
ELIZABETH MOUZON, *Administrative Secretary*

JOHN HURLEY, *Director, Board on Science and Technology for International Development*
MICHAEL MCD. DOW, *Associate Director, Studies*

Contents

INTRODUCTION 1
OVERVIEW 11

FOOD 17
    Introduction, 17
    Grains and Oilseeds, 18
    Tubers and Foliage, 26
    Leaf Protein, 28
    Fruits, 32
    Traditional Crops, 33
    References and Selected Readings, 39
    Research Contacts, 45

FUEL 50
    Introduction, 50
    Fuelwood Trees and Shrubs, 52
    Liquid Fuels, 65
    Gaseous Fuels, 67
    References and Selected Readings, 67
    Research Contacts, 72

FODDER 74
    Introduction, 74

    Grasses, 75
    Shrubs, 81
    Trees, 92
    References and Selected Readings, 95
    Research Contacts, 100

FIBER AND OTHER PRODUCTS 103
    Introduction, 103
    Essential Oils, 103
    Gums, Oils, and Resins, 105
    Pulp and Fiber, 109
    Bioactive Derivatives, 116
    Landscape and Ornamental Use, 120
    References and Selected Readings, 122
    Research Contacts, 127

INDEX 131
Board on Science and Technology for International
Development (BOSTID) 134
BOSTID Publications 135

Introduction

The agricultural use of saline water or soils can benefit many developing countries. Salt-tolerant plants can utilize land and water unsuitable for salt-sensitive crops (glycophytes) for the economic production of food, fodder, fuel, and other products. Halophytes (plants that grow in soils or waters containing significant amounts of inorganic salts) can harness saline resources that are generally neglected and are usually considered impediments rather than opportunities for development.

Salts occur naturally in all soils. Rain dissolves these salts, which are then swept through streams and rivers to the sea. Where rainfall is sparse or there is no quick route to the sea, some of this water evaporates and the dissolved salts become more concentrated. In arid areas, this can result in the formation of salt lakes or in brackish groundwater, salinized soil, or salt deposits.

There are three possible domains for the use of salt-tolerant plants in developing countries. These are:

1. Farmlands salinized by poor irrigation practices;
2. Arid areas that overlie reservoirs of brackish water; and
3. Coastal deserts.

In some developing regions, there are millions of hectares of salinized farmland resulting from poor irrigation practices. These lands would require large (and generally unavailable) amounts of water to leach away the salts before conventional crops could be grown. However, there may be useful salt-tolerant plants that can be grown on them without this intervention. Although the introduction of salt-tolerant plants will not necessarily restore the soil to the point that conventional crops can be grown, soil character is often improved and erosion reduced.

Moreover, many arid areas overlie saline aquifers—groundwater containing salt levels too high for the irrigation of conventional, salt-sensitive crops. Many of these barren lands can become productive by growing selected salt-tolerant crops and employing special cultural techniques using this store of brackish water for irrigation.

Throughout the developing world, there are extensive coastal deserts where seawater is the only water available. Although growing crops in sand and salty water is not a benign prospect for most farmers, for saline agriculture they can complement each other. The disadvantages of sand for conventional crops become advantages when saline water and salt-tolerant plants are used.

Sand is inherently low in the nutrients required for plant growth, has a high rate of water infiltration, and has low water-holding capacity. Therefore, agriculture on sand requires both irrigation and fertilizer. Surprisingly, 11 of the 13 mineral nutrients needed by plants are present in seawater in adequate concentrations for growing crops. In addition, the rapid infiltration of water through sand reduces salt buildup in the root zone when seawater is used for irrigation. The high aeration quality of sand is also valuable. This characteristic allows oxygen to reach the plant roots and facilitates growth. Although careful application of seawater and supplementary nutrients are necessary, the combination of sand, saltwater, sun, and salt-tolerant plants presents a valuable opportunity for many developing countries.

Of these three possibilities for the introduction of salt-tolerant plants (salinized farmland, undeveloped barren land, and coastal deserts), the reclamation of degraded farmland has several advantages: people, equipment, buildings, roads, and services are usually present and a social structure and market system already exist. The potential use of saline aquifers beneath barren lands depends on both the concentration and nature of the salts. The direct use of seawater for agriculture is probably the most challenging potential application.

Most contemporary crops have been developed through the domestication of plants from nonsaline environments. This is unfortunate since most of of the earth's water resources are too salty to grow them. From experience in irrigated agriculture, Miyamoto (personal communication) suggests the following classification of potential crop damage from increasing salt levels:

Irrigation Water	Salts, ppm	Crop Problems
Fresh	<125	None
Slightly saline	125-250	Rare
Moderately saline	250-500	Occasional
Saline	500-2,500	Common
Highly saline	2,500-5,000	Severe

Colorado River water, used for irrigation in the western United States, contains about 850 ppm of salts; seawater typically contains 32,000-36,000 ppm of salts. Salinity levels are usually expressed in terms of the electrical conductivity (EC) of the irrigation water or an aqueous extract of the soil; the higher the salt level, the greater the conductivity. The salinity of some typical water sources is shown in Table 1.

TABLE 1 Water Salinity.

Salinity Measurement	Irrigation Water Quality (Good)	(Marginal)	Colorado River	Alamo River	Negev Groundwater	Pacific Ocean
Electrical conductivity (dS/m)*	0-1	1-3	1.3	4.0	4.0 - 7.0	46
Dissolved solids, ppm	0-500	500-1,500	850	3,000	3,000-4,500	35,000

*1 dS/m = 1 mmho/cm = (approx.) 0.06%NaCl = (approx.) 0.01 mole/l NaCl. 10,000 ppm = 10 o/oo (parts per thousand) = 10 grams per liter = 1.0%

In the International System of Units (SI), the unit of conductivity is the Siemens symbol, S, per meter. The equivalent unit commonly appearing in the literature is the mho (reciprocal ohm); 1 mho equals 1 Siemen.

SOURCE: Adapted from Epstein, 1983; Pasternak and De Malach, 1987; and Rhoades et al., 1988.

There are three broad approaches to utilizing saline water, depending on the salt levels present. These include the use of marginal to poor irrigation water with electrical conductivities (ECs) up to about 4 dS/m, the use of saline groundwaters such as those in Israel's Negev Desert with ECs up to about 8 dS/m, and the use of even more saline waters with salt concentrations up to that of seawater.

At low, but potentially damaging, salt levels, Rhoades and coworkers (1988) have grown commercial crops without the yield losses that would normally be anticipated. Through knowledge of crop sensitivity to salt at various growth stages, they used combinations of Colorado River water and Alamo River water to minimize the use of the higher quality water. For example, wheat seedlings were established with Colorado River water; Alamo River water was then used for irrigation through harvest with no loss in yield.

At higher salt levels, Pasternak and coworkers (1985) have developed approaches that involve special breeding and selection of crops and meticulous water control. The agriculture of Negev settlements in Israel is based on the production of cotton with higher yields, quality tomatoes for the canning industry, and quality melons for export—all grown with EC 4-7 dS/m groundwater. Experimental yields of a wide variety of traditional crops grown in Israel with water with ECs up to 15 dS/m, are shown in Table 6 (p. 35). In west Texas (USA), Miyamoto and coworkers (1984) report commercial production of alfalfa, melons, and tomatoes with EC 3-5 dS/m irrigation water, and cotton with 8 dS/m irrigation water.

The use of water with still higher salt levels up to, including, and even exceeding that of seawater for irrigation of various food, fuel, and fodder crops has been reported by many researchers including Aronson (1985; 1989), Boyko (1966), Epstein (1983; 1985), Gallagher (1985), Glenn and O'Leary (1985), Iyengar (1982), Pasternak (1987), Somers (1975), Yensen (1988), and others. These scientists have produced grains and oilseeds; grass, tree, and shrub fodder; tree and shrub fuelwood; and a variety of fiber, pharmaceutical, and other products using highly saline water.

Thus, depending on the soil or water salinity levels, salt-tolerant plants can be identified that will perform well in many environments in developing countries. The salt tolerance of some of these plants enables them to produce yields under saline conditions that are comparable to those obtained from salt-sensitive crops grown under nonsaline conditions.

The maximum amount and kind of salt that can be tolerated

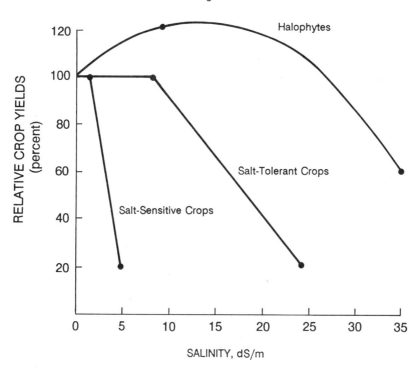

FIGURE 1 Growth response to salinity. Many halophytes, such as *Suaeda maritima*, have increased yields at low salinity levels. Salt-tolerant crops, such as barley, maintain yields at low salinity levels but decrease as salt levels exceed a certain limit. Yields of salt-sensitive crops, such as beans, decrease sharply even in the presence of low levels of salt. SOURCE: Adapted from Greenway and Munns, 1980; Maas 1986; and Yensen, et al., 1985.

by halophytes and other salt-tolerant plants varies among species and even varieties of species. Many halophytes have a special and distinguishing feature—their growth is improved by low levels of salt. Other salt-tolerant plants grow well at low salt levels but beyond a certain level growth is reduced. With salt-sensitive plants, each increment of salt decreases their yield (Figure 1).

Such data provide only relative guidelines for predicting yields of crops grown under saline conditions. Absolute yields are subject to numerous agricultural and environmental effects. Interactions between salinity and various soil, water, and climatic conditions all affect the plant's ability to tolerate salt. Some halophytes require fresh water for germination and early growth but can tolerate higher salt levels during later vegetative and reproductive stages. Some can germinate at high salinities but require lower salinity for maximal growth.

Traditional farming efforts usually focus on modifying the environment to suit the crop. In saline agriculture, an alternative is to allow the environment to select the crops, to match salt-tolerant plants with desirable characteristics to the available saline resources.

In many developing countries extensive areas of degraded and arid land are publicly owned and readily accessible for government-sponsored projects. These lands are often located in areas of high nutritional and economic need as well. If saline water is available, the introduction of salt-tolerant plants in these regions can improve food or fuel supplies, increase employment, help stem desertification, and contribute to soil reclamation.

LIMITATIONS

Undomesticated salt-tolerant plants usually have poor agronomic qualities such as wide variations in germination and maturation. Salt-tolerant grasses and grains are subject to seed shattering and lodging. The foliage of salt-tolerant plants may not be suitable for fodder because of its high salt content. Nutritional characteristics or even potential toxicities have not been established for many edible salt-tolerant plants. When saline irrigation water is used for crop production, careful control is necessary to avoid salt buildup in the soil and to prevent possible contamination of freshwater aquifers.

Most importantly, salt-tolerant plants should not be cultivated as a substitute for good agricultural practice nor should they be used as a palliative for improper irrigation. They should be introduced only when and where conventional crops cannot be grown. Also, currently productive coastal areas (such as mangrove forests) should be managed and restored, not converted to other uses.

All of these limitations are impediments to the use of conventional methods for culture and harvest of salt-tolerant plants and the estimation of their production economics.

RESEARCH NEEDS

Increased research on the development of salt-tolerant cultivars of crop species could, with appropriate management, result in the broader use of saline soils. In the early selection and breeding programs of crop species for use in nonsaline environments, performance was improved through the considerable genetic variability present in the unimproved crops and in their wild relatives. Since few crops have been subjected to selection for salinity tolerance, it is possible that variation in this characteristic may also exist. Conversely, few undomesticated salt-tolerant plants have been examined for variability in their agronomic qualities, and it is even more likely that such characteristics can be improved through breeding programs.

In addition, germplasm collection and classification, breeding and selection, and development of cultural, harvest, and postharvest techniques are all needed. Basic information on the way in which plants adapt to salinity would significantly assist their economic development.

Exploration for new species should continue to identify candidates for economic development. Research can then begin on ways to improve the agronomic qualities of these plants and to utilize their genetic traits. For example, seed from a wild tomato found on the seashore of the Galapagos Islands produced tomatoes that were small and bitter. When this species was crossed with a commercial tomato cultivar, flavorful fruit the size and color of cherry tomatoes were obtained in 70 percent seawater.

Recent advances in plant biotechnology include work on salinity tolerance and productivity. New techniques for *in vitro* selection of genotypes tolerant to high salinity levels have been found to improve the adaptability of conventional crops as well as assist in the selection of desired genotypes from a wide range of natural variability in individual salt-tolerant plants. Genotypes with increased tolerance to water and salinity stress have been identified and followed in genetic crosses with conventional genotypes using new techniques in gene mapping and cell physiology.

Stress genes are now the target of research in genetic engineering. The transfer of these genes from sources in salt-tolerant species to more productive crops will require modifications in cultural practices as well as treatment of the plant products.

Interdisciplinary communication is particularly important in research on salt- tolerant plants. Cooperation among plant ecologists, plant physiologists, plant breeders, soil scientists, and agricultural engineers could accelerate development of economic crops. Further, universities could introduce special programs to allow broad study of the special characteristics of saline agriculture to serve growing needs in this field.

REFERENCES AND SELECTED READINGS

Abrol, I. P., J. S. P. Yadav and F. I. Massoud. 1988. *Salt-affected Soils and Their Management.* Soils Bulletin 39, FAO, Rome, Italy.

Ahmad, R. 1987. *Saline Agriculture at Coastal Sandy Belt.* University of Karachi, Karachi, Pakistan.

Ahmad, R. and A. San Pietro (eds.). 1986. *Prospects for Biosaline Research.* University of Karachi, Karachi, Pakistan.

Aronson, J. A. 1989. *Haloph: Salt-tolerant Plants of the World.* University of Arizona, Tucson, Arizona, US.

Aronson, J. A. 1985. Economic halophytes—a global view. Pp. 177-188 in: G. E. Wickens, J. R. Goodin and D. V. Field (eds.) *Plants for Arid Lands.* George Allen and Unwin, London, UK.

Bahri, A. 1987. Utilization of saline waters and soils in Tunisia. Results and research prospects. *Fertilizers and Agriculture* 96:17-34.

Barrett-Lennard, E. G., C. V. Malcolm, W. R. Stern and S. M. Wilkins (eds.). 1986. *Forage and Fuel Production from Salt Affected Wasteland.* Elsevier Publishers, Amsterdam, Netherlands.

Bernstein, L. 1964. *Salt Tolerance of Plants.* USDA Bulletin No. 283, Washington, DC, US.

Boyko, H. 1966. *Salinity and Aridity. New Approaches to Old Problems.* Dr. W. Junk, Publisher, The Hague, Netherlands.

Epstein, E. 1985. Salt tolerant crops: origins, development, and prospects of the concept. *Plant and Soil* 89:187-198.

Epstein, E. 1983. Crops tolerant of salinity and other stresses. Pp. 61-82 in: *Better Crops for Food.* Pitman Books, London, UK.

Epstein, E., J. D. Norlyn, D. W. Rush, R. W. Kingsbury, D. B. Kelley, G. A. Cunningham and A. F. Wrona. 1980. Saline culture of crops: a genetic approach. *Science* 210:399-404.

Flowers, T. J., M. A. Hajibagheri and N. J. W. Clipson. 1986. Halophytes. *The Quarterly Review of Biology* 61(3):313-337.

Gallagher, J. L. 1985. Halophytic crops for cultivation at seawater salinity. *Plant and Soil* 89:323-336.

Glenn, E. P. and J. W. O'Leary. 1985. Productivity and irrigation requirements of halophytes grown with seawater in the Sonoran Desert. *Journal of Arid Environments* 9(1):81-91.

Goodin, J. R. and D. K. Northington. 1979. *Arid Land Plant Resources.* Texas Tech University, Lubbock, Texas, US.

Greenway, H. and R. Munns. 1980. Mechanisms of salt tolerance in nonhalophytes. *Annual Review of Plant Physiology* 31:149-190.

Iyengar, E. E. R. 1982. Research on seawater irriculture in India. Pp. 165-175 in: A. San Pietro (ed.) *Biosaline Research: A Look to the Future.* Plenum Press, New York, New York, US.

Maas, E. V. 1986. Crop tolerance to saline soil and water. Pp. 205-219 in: R. Ahmad and A. San Pietro (eds.) *Prospects for Biosaline Research.* University of Karachi, Karachi, Pakistan.

Miyamoto, S., J. Moore and C. Stichler. 1984. Overview of saline water irrigation in far west Texas. Pp. 222-230. in: *Proceedings of Irrigation and Drainage Speciality Conference,* ASCE, Flagstaff, Arizona, July 24-26, 1984.

Mudie, P. J. 1974. The potential economic uses of halophytes. Pp. 565-597 in: R. J. Reimold and W. H. Queen (eds.) *Ecology of Halophytes*. Academic Press, New York, New York, US.

O'Leary, J. W. 1985. Saltwater crops. *CHEMTECH* 15(9):562-566.

Pasternak, D. and Y. De Malach 1987. Saline water irrigation in the Negev Desert. in: *Proceedings: Agriculture and Food Production in the Middle East*. Athens, Greece. January 21-26, 1987.

Pasternak, D. 1987. Salt tolerance and crop production—a comprehensive approach. *Annual Review of Phytopathology* 25:271-291.

Pasternak, D. and A. San Pietro (eds.). 1985. *Biosalinity in Action: Bioproduction with Saline Water*. Martinus Nijhoff Publishers, Dordrecht, The Netherlands

Pasternak, D., A. Danon and J. A. Aronson. 1985. Developing the seawater agriculture concept. *Plant and Soil* 89:337-348.

Raz, B., S. Dover and E. Udler. 1987. Desert agriculture. *Science and Public Policy* 14(4):207-216.

Rhoades, J. D., F. T. Bingham, J. Letey, A. R. Dedrick, M. Bean, G. J. Hoffman, W. J. Alves, R. V. Swain, P. G. Pacheco and R. D. Lemert. 1988. Reuse of drainage water for irrigation: results of Imperial Valley study. *Hilgardia* 56(5):1-44.

Rick, C. M. 1972. Potential genetic resources in tomato species: clues from observations in native habitats. Pp. 255-269 in: A. M. Srb (ed.) *Genes, Enzymes, and Populations*. Plenum Press, New York, New York, US.

San Pietro, A. (ed.). 1982. *Biosaline Research. A Look to the Future*. Plenum Press, New York, New York, US.

Shainberg, I. and J. Shalhevet (eds.). 1984. *Soil Salinity under Irrigation. Processes and Management*. Springer-Verlag, New York, New York, US.

Sharma, S. K. and I. C. Gupta. 1986. *Saline Environment and Plant Growth*. Agro Botanical Publishers, Bikaner, India.

Somers, G. F. (ed.). 1975. *Seed-bearing Halophytes as Food Plants*. Proceedings of a conference. College of Marine Studies, University of Delaware, Lewes, Delaware, US.

Staples, R. C. and G. H. Toenniessen (eds.). 1984. *Salinity Tolerance in Plants*. Wiley-Interscience, New York, New York, US.

United States Department of Agriculture. 1954. *Diagnosis and Improvement of Saline and Alkali Soils*. USDA Handbook 60. USDA, Washington, DC, US.

Yensen, N. P., S. B. Yensen and C. W. Weber. 1985. A review of *Distichlis* spp. for production and nutritional values. Pp. 809-822 in: E. E. Whitehead, C. F. Hutchinson, B. N. Timmermann, and R. G. Varady (eds.) *Arid Lands Today and Tomorrow*, Westview Press, Boulder, Colorado, US.

Yensen, N. P. 1988. Plants for salty soil. *Arid Lands Newsletter* 27:3-10. University of Arizona, Tucson, Arizona, US.

Whitehead, E. E., C. F. Hutchinson, B. N. Timmermann and R. G. Varady (eds.). 1985. *Arid Lands Today and Tomorrow*. Westview Press, Boulder, Colorado, US.

Wickens, G. E., J. R. Goodin and D. V. Field (eds.). 1985. *Plants for Arid Lands*. George Allen & Unwin, London, UK.

Overview

Scientists exploring seashores, estuaries, and saline seeps have found thousands of halophytes with potential use as food, fuel, fodder, fiber, and other products. Many have already been in traditional use, and there are also a number of plants that, although not halophytes, have sufficient salt tolerance for use in some saline environments.

Although economic consideration of halophytes and other salt-tolerant plants is just beginning, they are now receiving increased attention in arid regions where intensive irrigation has led to salinized soils or where water shortages are forcing use of marginal resources such as brackish underground water. This report will examine some of the plants that may be suitable for economic production in saline environments in developing countries.

There are four sections in this report. They highlight salt-tolerant plants that may serve as food, fuel, fodder, and other products such as essential oils, pharmaceuticals, and fiber. In each of these sections, plants are described that have potential for productive use. Each section also contains an extensive list of recent papers and other publications that contain additional information on these plants. A list of researchers currently working on these plants or related projects is included at the end of each section.

FOOD

Many halophytes survive saline stress by accumulating salt in their vegetative tissues. The salt levels in the leaves and stems of these plants can limit their direct consumption as food, but their seeds are relatively salt-free, which may allow production of starchy grains or oilseeds.

For example, the seeds of *Zostera marina*, a sea grass, were used as food by the Seri Indians of the southwestern United States; in recent tests, these seeds were ground to flour and used to make bread. Seeds of *Distichlis palmeri*, Palmer's saltgrass, were harvested from tidal flats at the head of the Gulf of California by Cocopa Indians. The seed, about the same size as wheat, has also been used for making bread.

The production of vegetable oils from seed-bearing halophytes appears promising. A number of these seeds have an oil content comparable with that of better known sources of vegetable oils. A *Salicornia* species is being evaluated as a source of vegetable oil in field trials in the United Arab Emirates, Kuwait, and Egypt. Since many developing countries import vegetable oils, the opportunity for domestic production on currently unusable lands warrants investigation.

It may be possible as well to use the salt-containing vegetative parts of some halophytes to produce salt-free leaf protein. In this process, any inorganic salts in the leaves are separated from the protein. Leaf protein production may help improve the nutritional quality of foods in developing countries.

There are also traditional food crops that are grown commercially using underground brackish water for irrigation. These include tomatoes, onions, and melons. Asparagus also appears to grow well with brackish water irrigation.

FUEL

More than a billion people in developing countries rely on wood for cooking and heating. In most developing areas, the rate of deforestation for fuelwood and for agricultural expansion far exceeds the rate of reforestation. Increasing needs for agricultural land to feed growing populations make it unlikely that land suitable for food crops will be used for tree planting. One alternative, therefore, is to use marginal or degraded lands to produce more fuelwood.

Fuelwood and building materials can be produced from salt-tolerant trees and shrubs employing land and water unsuitable for conventional crops. Fuel plantations established on saline soils or irrigated with saline water would allow more fertile land and fresh water to be reserved for food or forage production. With careful planning, trees and shrubs can help rehabilitate degraded lands by stabilizing the ecosystem and providing niches and protection for other plants and animals.

In Australia, a consortium of business and academic groups is developing a program to market salt-tolerant trees for fuel and pulp. The project will screen Australian tree species for growth rates, salt tolerance, and drought tolerance. Root fungi associated with these trees, which help the trees obtain nutrients from the soil, will be screened for salt tolerance and their influence on tree growth. Trees with superior growth on saline soils will be tissue cultured and inoculated with salt-tolerant root fungi. These cloned trees will then be tested for field performance in Australia and developing countries.

FODDER

Halophytic grasses, shrubs, and trees are all potential sources of fodder. Trees and shrubs can be valuable components of grazing lands and serve as complementary nutrient sources to grasses in arid and semiarid areas.

Among the grasses, kallar grass (*Leptochloa fusca*) tolerates waterlogging and recovers well from cutting and grazing. Its economic value as fodder for buffalo and goats has already been demonstrated in Pakistan and is now being examined in other countries. Members of the *Spartina* genus (cordgrasses) have also been used as fodder. These tough, long-leaved grasses are found in tidal marshes in North America, Europe, and Africa. The salt-tolerant grass *Sporobolus virginicus* has also been used as cattle forage.

Distichlis spicata has been used as forage for cattle in Mexico. Introduced in the area of a dry salt lake outside Mexico City, *D. spicata* reduced windblown dust while serving as cattle feed.

In arid and semiarid zones, trees and shrubs for fodder have several advantages over grasses. They are generally less susceptible than grasses to fire and to seasonal variation in moisture availability and temperature. Usually less palatable than grasses, they can provide reserve or supplementary feed sources.

Among the shrubs, saltbushes (*Atriplex* spp.) grow throughout the world. They tolerate salinity in soil and water, and many are perennial shrubs that remain green all year. They are especially useful as forage in arid zones.

Among trees, *Acacia* species are widely used in arid and saline environments as supplementary sources of fodder. *Acacia* pods provide food for livestock in large areas of the semiarid zone of Africa. *Acacia cyclops* and *A. bivenosa* tolerate salt spray and salinity. They grow on coastal dunes as small trees or bushy shrubs. Pods and leaves of both are consumed by goats.

Leucaena leucocephala is a tree legume widely cultivated in tropical and subtropical countries. Leaves, pods, and seeds are grazed by cattle, sheep, and goats. In Pakistan, it has been grown on coastal sandy soil through irrigation with saline water. When seawater comprised 20 percent of the irrigation water, yields were reduced by 50 percent, however.

The leaves and pods of mesquite (*Prosopis* spp.) have been used as forage for cattle, goats, sheep, and camels in countries throughout the world—*P. juliflora* and *P. cineraria* in India, *P. chilensis* in South America, *P. glandulosa* in the United States, and *P. pallida* in Australia.

About 20 years ago the Chilean government began to improve the salt-afflicted Pampa del Tamarugal in the northern part of the country by growing tamarugo (*P. tamarugo*). In some cases, these trees were planted in pits dug through the salt into the soil. Although watering was required for the first year, after that the plants survived by capturing moisture from the ground and air. About 23,000 hectares are now covered with tamarugo forest. The tamarugo leaves and fruit are used as feed for sheep and goats.

FIBER AND OTHER PRODUCTS

Salt-tolerant plants can also be used to produce economically important materials such as essential oils, flavors, fragrances, gums, resins, oils, pharmaceuticals, and fibers. They may also be marketed for use in landscape gardening, and for their foliage or flowers.

In India, peppermint oil and menthol have been produced in saline environments. The salt-tolerant *kewda*, a common species of screwpine, is used to produce perfume and flavoring ingredients.

Sesbania bispinosa, commonly known as *dhaincha* in India, is an important salt-tolerant legume and fodder crop. In addition to use of the stalks as sources of fiber and fuel, the seeds yield a galactomannan gum that can be used for sizing and stabilizing applications, and a seed meal that can be used for poultry and cattle feed.

Grindelia camporum is a salt-tolerant resinous perennial shrub. It produces large amounts of aromatic resins that have properties similar to the terpenoids in wood and gum rosins, which are used commercially in adhesives, varnishes, paper sizings, printing inks, soaps, and numerous other industrial applications.

Jojoba (*Simmondsia chinensis*) is a perennial desert shrub with seeds that contain a unique oil similar to that obtained from the sperm whale. This oil and its derivatives have been used primarily in cosmetics, but broader use in lubricants and waxes will probably develop if prices come down. Jojoba is relatively salt tolerant. In Israel, jojoba is growing well near the Dead Sea with brackish water irrigation.

Phragmites australis, common reed, is an ancient marsh plant that has served in roofing, thatching, basketmaking and fencing, as well as being used for fuel. It grows throughout the world in water-saturated soils or standing waters that are fresh or moderately saline. In Egypt, two salt-tolerant rushes, *Juncus rigidus* and *J. acutus*, have been investigated with particular emphasis on their potential use in papermaking.

Many attractive halophytes can be used as landscape plants, especially in areas with constraints on the use of fresh water for watering or irrigation. In Israel, salt-tolerant trees and shrubs are sold for amenity planting. In addition, other salt-tolerant plants have potential for cut-flower production.

Although the salt-tolerant plants described in this report typify those that are currently being evaluated or appear to deserve additional attention, the inventory is far from complete. Many other species may have equal or greater potential. In some cases in this report, specific companies or products are identified. This is for convenience and does not constitute endorsement.

1
Food

INTRODUCTION

Saline agriculture can provide food in several ways. Appropriate salt-tolerant plants currently growing in saline soil or water can be domesticated and their seeds, fruits, roots, or foliage used as food. When the foliage is too high in salt for direct consumption, the leaves can be processed to yield salt-free protein, which can be used to fortify traditional foods. In addition, conventional food crops can be bred or selected to tolerate mildly saline water.

This section will examine some of the little-known seed-bearing plants that grow in saline environments and their special characteristics, the use of foliage from salt-tolerant plants to produce leaf protein, some salt-tolerant fruits, and the performance of some conventional food crops with saline water.

Of conventional crops, the only ones with halophytic ancestors are sugar-, fodder-, and culinary beets (all *Beta vulgaris*) and the date palm (*Phoenix dactylifera*). These plants can be irrigated with brackish water without serious loss of yield. Of about 5,000 crops that are cultivated throughout the world, few can survive with water that contains more than about 0.5 percent salt, and most suffer serious losses of yield at about 0.1 percent salt. In searching for crops for saline agriculture, those that currently comprise the bulk of human food should be considered as models—maize, wheat, rice, potatoes, and barley. If these major crops can be grown using saline resources, or if new, salt-tolerant crops that are acceptable substitutes can be developed, the world's food supply will have a more diverse and vastly expanded base.

Along with significant technical impediments to the widespread use of saline resources for food production, social barriers may exist as well. Food preparation is one of mankind's most culture-bound activities. Food selection, cooking method and participants, flavor, consistency, and serving time and place are often established by long tradition, and practitioners are resistant to change. New foods that require significant changes in any of these practices are unlikely to be readily accepted.

GRAINS AND OILSEEDS

Many seed-bearing halophytes have an interesting characteristic: although they may have significantly greater levels of salt in their stems, branches, and leaves than conventional plants, their seeds are relatively salt-free. Seeds of halophytes and salt-sensitive plants have about the same ash and salt content, as shown in Table 2.

TABLE 2 Protein, Oil, and Ash Contents of Seeds from Salt-Sensitive and Salt-Tolerant Plants.

Seed	Percent of Dry Weight as		
	Protein	Oil	Ash
Salt-Sensitive			
Safflower	14.3	30.4	2.5
Sesame	18.6	49.1	5.3
Soybean	40.0	18.8	4.8
Sunflower	17.5	36.0	3.6
Salt-Tolerant			
Atriplex canescens	5.4	1.0	6.5
Atriplex triangularis	16.4	9.4	3.5
Cakile edentula	28.6	52.2	5.2
Cakile maritima	21.5	47.1	5.0
Chenopodium quinoa	12.1	7.5	3.1
Crithmum maritimum	21.5	41.4	8.0
Kosteletzkya virginica	23.8	18.1	5.0

SOURCE: O'Leary, 1985.

This has valuable consequences. Although the direct consumption of halophyte vegetative tissue by humans and animals can be limited by its salt content, the seeds of many halophytes present no such obstacle. This allows consideration of a wide variety of seed-producing halophytes as new sources of grains or vegetable oils.

Some salt-tolerant grains and oilseeds have already been used or examined.

Almost fifty species of seed-bearing seagrasses grow in nearshore areas of the world's oceans. One of these, *Zostera marina*, grows fully submerged in seawater.

Eelgrass (*Zostera marina*) grows well in the Gulf of California in North America. In this region, seawater temperatures seldom fall below 12°C and can reach 32°C in summer. Sunlight is intense. At maturity in the spring, the reproductive stems bearing the seed break loose and are washed ashore. Harvest involves collecting these stems and separating the seeds. The seeds, 3-3.5 mm long and weighing up to about 5.6 mg, contain about 50 percent starch, 13 percent protein, and 1 percent fat. The Seri Indians used this seed as one of their major foods.

Although the potential for growing a food crop directly in seawater is attractive, there are obstacles to broader cultivation of eelgrass. Coastal deserts offer the best possibility, but tidal action is required; these grasses apparently cannot grow in stagnant water. In warm, dry climates the plants can tolerate only short exposure to the air.

Palmer saltgrass (*Distichlis palmeri*) grows in tidal flats and marshy inlets in the Gulf of California, and thrives with tidal inundations of seawater. It is a perennial with tough rhizomes from which emerge densely crowded stems about 0.5 m tall. The spikelets, which bear the seed, readily shatter and are also dislodged by tidal action. Although this shattering is generally undesirable in a crop (because seed on the ground is difficult to gather), with Palmer saltgrass, the spikelets float and are washed ashore. These seeds were gathered by the Yuman Indians, ground into flour, and consumed as a gruel. It can also be used to make bread.

Once established, Palmer saltgrass should not need replanting. Preliminary observations indicate that it is fast-growing and the standing crop is extremely dense. These dense stands along with the saline conditions should reduce competition from weeds. Field tests with hybrid cultivars of this crop yielded about 1,000 kg of grain per hectare when irrigated with water containing 1-3 percent salt. Optimum yields are projected to be obtained at about 2 percent

TABLE 3 Nutritional Composition of *Distichlis palmeri* vs. Wheat and Barley.

Crop	Percent of				
	Protein	Fiber	Fat	Ash	Carbohydrate
D. palmeri	8.7	8.4	1.8	1.6	79.5
Wheat	13.7	2.6	1.9	1.9	79.9
Barley	13.0	6.0	1.9	3.4	75.7

SOURCE: Yensen, 1985.

salinity. The nutritional characteristics of *D. palmeri* are summarized in Table 3.

The grain from a *D. palmeri* variety developed by NyPa, Inc. has a well-balanced amino acid profile and three times the fiber of common wheat. Antinutritional phytic acid is very low, and gluten, a potentially allergenic protein, is not present in detectable amounts.

Alkali sacaton (*Sporobolus airoides*) is a widespread perennial grass in the western United States and northern Mexico, often occurring on alkaline or semisaline soils. Its 0.95-1.2 mm grain is edible and was probably a significant food resource for Hopi and Paiute Indians of the North American Southwest. The grain is readily separated, produced in large quantity, and should be suitable for harvesting with a basket. Although *S. helvolus* and *S. maderaspatanus* also grow on saline soils, the use of their grain as food has not been reported.

Pearl millet or bajra (*Pennisetum typhoides*), a popular food grain in Africa and India, has been grown on coastal dunes near Bhavnagar using seawater (EC = 26.6-37.5 dS/m) for irrigation. When seedlings were established with fresh water and fertilizer applied, multiple irrigations with seawater gave yields of 1.0-1.6 tons per hectare of grain and 3.3-6.5 tons per hectare of fodder.

Quinoa (*Chenopodium quinoa*) is a staple of the Andean highlands. An annual herb, quinoa grows 1-2.5 m tall at altitudes of 2,500-4,000 m. The plant matures in 5-6 months, producing white or pink seeds in large sorghum-like clusters. Although the seeds are small, they comprise 30 percent of the dry weight of the plant. Yields of 2,500 kg per hectare have been reported. Quinoa has a protein content that is higher, and an amino acid composition that is better balanced, than the major cereals. Although quinoa has bitter tasting constituents—chiefly saponins—in the seed's outer layer, these can

Seawater has been used for the irrigation of bajra, a popular millet in India. After seedlings were established with fresh water, multiple irrigations with seawater gave yields of up to 1.6 tons per hectare of grain. (E.R.R. Iyengar)

be removed by washing the seeds in cold water. The seeds are traditionally used in soup or ground into flour for bread and cake. They have also been used for brewing beer and for animal feed.

Somers (1982) reported that quinoa germinated in a mixture of one-third seawater and two-thirds fresh water but would not grow at this salinity. In the salt flats of southern Bolivia and northern Chile, quinoa is one of the few crop plants grown. In this arid region (230 mm annual rainfall), quinoa is planted in holes about 40 cm deep where the soil is damp. As the plant grows, soil is filled in around it. With wide stretches of salt beds nearby, the environment is certainly saline, but no measurements have been reported.

Seashore mallow (*Kosteletzkya virginica*) is a perennial surviving about five years in cultivation. Although the seeds must be germinated at low salinity, the plant can tolerate 2.0-2.5 percent salinity during growth. Hulled seeds, which resemble millet, contain as much as 32 percent protein and 22 percent oil. Grain yields from plots irrigated with water containing 2.5 percent salt have ranged from 0.8 to 1.5 tons per hectare.

Quinoa is a primary grain of the Andean highlands. Although precise salt tolerance has not been established, quinoa commonly grows near salt flats. (L. McIntyre)

TABLE 4 Acacia Seed Composition.

Species	Energy (kJ)	Protein	Fat	Percentage of Carbohydrate	Water	Ash
A. aneura	2220	23.3	37.0	25.5	4.3	9.7
A. coriacea	1491	23.8	7.7	48.1	17.1	3.7
A. cowleana	1507	22.2	10.1	44.6	15.6	7.2
A. dictyophleba	1519	26.8	6.3	49.0	11.2	6.5
Wheat*		13.7	1.9	79.9	--	1.9

*Water-free composition
SOURCE: Peterson, 1978.

Many *Acacia* seeds are rich in nutrients with higher energy, protein, and fat contents than wheat or rice. The high protein levels (~20 percent) suggest breadmaking potential, and the high fat contents (up to 37 percent) indicate potential as oilseeds.

About 50 of the 800 species of *Acacia* found in Australia have been used as food by Australian aborigines. Twenty of these were staple foods. In most cases dry ripe seeds were ground to a coarse flour that was then mixed with water to give an unleavened dough, which was baked on hot stones or in the ashes of a fire. Table 4 provides some information on a few *Acacia* seeds.

Seeds from salt-tolerant *Tecticornia* species were also used by Australian aborigines. The small (1.5-1.8 mm) seeds were ground to flour and used for making bread. *T. australasica* and *T. verrucosa* grow to about 40 cm in coastal mudflats above the normal tidal level. Germination of the seed appears to be dependent on seasonal rains leaching the salt from the upper soil layer. *T. verrucosa* also occurs inland on moderately saline flats.

Indian almond (*Terminalia catappa*) is an erect tree reaching 15-25 m. It probably originated in Malaysia and was spread by its fruits carried on ocean currents. It is cultivated in much of India and Burma and has become common in east and west Africa, the Pacific Islands, and in coastal areas of tropical America. Its ellipsoidal fruit is 4-7 cm long and 2.5-3.8 cm wide, the edible kernel is 3-4 cm long and 3-5 mm thick, and, in many varieties, the fruit is sweet and palatable. The nut is used as an almond substitute, and the wood is valued for construction and furniture use. The tree seems well adapted to sandy and rocky coasts. In Florida, it is known to withstand flooding, wind, and ocean spray, as well as saline soils.

The salt-tolerant *Tecticornia verrucosa* grows to about 40 cm in coastal mudflats above the normal tidal level. Seeds from this strange-looking plant were used by Australian aborigines for making bread. (P. Wilson)

The Indian almond tree is grown for its durable lumber as well as for its edible nuts. It is reported to withstand flooding, wind, and ocean spray, as well as saline soils. (J. Morton)

Argan (*Argania spinosa*) covers an area of about 600,000 hectares of bushland in southwest Morocco. It can develop as a shrub or tree, usually in dense clumps. It has an important role as a browse and an edible oil is produced from its seeds. Preliminary work to determine its salt tolerance has been initiated in Israel.

A *Salicornia* species, described as SOS-7, has been grown in field trials in Mexico, Egypt, and the United Arab Emirates to produce an edible, safflower-like seed oil. When irrigated with seawater, about 20 tons of plant material per hectare are obtained. The oilseeds comprise about 2 tons of this total. The straw can be used for about 10 percent of the feed for cattle, goats, and sheep.

Prior to planting this *Salicornia* on salt flats near Kalba, United Arab Emirates, the soil was leached with seawater to reduce the salt level. *Salicornia* was then grown with seawater irrigation, and used to feed Damascus goats. The researchers estimate that one hectare of *Salicornia* could raise up to twenty goats or sheep.

Argan covers a wide area of bushland in southwest Morocco. It has an important role as a browse and an edible oil is produced from its seeds. (G. Voss)

TUBERS AND FOLIAGE

Wild water chestnut (*Eleocharis dulcis*) occurs in saline coastal swamps in Southeast Asia and Oceania. The tubers, smaller and harder than those of superior varieties cultivated in fresh water, are traditionally gathered from shallow waters and cooked as delicacies or pounded to meal.

The roots and stems of saltwort (*Batis maritima*) were used as food by the Seri Indians in the southwestern United States. Using seawater irrigation, dry weight yields of 17 tons per hectare have been obtained.

Seaside purslane (*Sesuvium portulacastrum*) is a wide-spreading, succulent, perennial herb valued as an edible wild plant in tropical coastal areas of the United States and the Caribbean. It is cultivated and consumed as a vegetable in India, Indonesia, and southern China. Boiling with several changes of water is necessary to eliminate excess salt. Analysis of the edible portion shows high values for calcium, iron, and carotene. In India, it is also used as fodder.

Common purslane (*Portulaca oleracea*) is also used as a potherb

At Kino Bay, Mexico, a *Salicornia* species is mechanically harvested. Yields of about 2 tons per hectare of oilseed and 18 tons per hectare of fodder are obtained with seawater irrigation. (H. Weiss)

and in salads and soups. It is reported to contain 29 mg per 100 g of vitamin C and a vitamin A potency of 7,500 I.U. per 100 g.

The leaves of sea fennel (*Crithmum maritimum*) have been used as a medicinal herb, a spice, and as a salad ingredient. They contain significant amounts of vitamin C and have traditional use in protecting sailors from scurvy. About 100 g of fresh leaves provides the recommended daily allowance of vitamin C.

The leaves of *Atriplex triangularis* are similar to spinach in appearance and nutritional composition. It is a leafy annual vegetable that grows on the edge of coastal marshes in eastern North America. Selection among collected lines at the University of Delaware has resulted in a cultivar that gives an estimated yield of 21,300 kg per hectare (fresh weight) using seawater for irrigation. *A. hortensis* is also cultivated in India for its spinach-like leaves.

The ice plant (*Mesembryanthemum crystallinum*) is native to South Africa. A succulent annual herb, it grows on sea coasts and salty deserts. The leaves and seeds of the plant are reported to be edible.

Sea fennel leaves contain vitamin C and have traditional use in protecting sailors from scurvy. This plant is growing with seawater irrigation near Ashkelon, Israel. (G. Shay)

Common Indian saltwort (*Suaeda maritima*) occurs in saline soils along the eastern and western coasts of India. It has been used for fixing seashore sand dunes. Its green leaves are considered a wholesome vegetable.

Leaf Protein

Although the leaves and shoots of some salt-tolerant foliage crops can be used in salads or as a garnish with minimal processing, most halophytes retain enough salt in their leaves to inhibit their consumption. One solution to this problem is to extract leaf protein from the salt-containing foliage.

To produce leaf protein, fresh foliage is fed into a press and the juice extracted. The fibrous material remaining after the juice is extracted from the leaves can be used as ruminant feed. The juice is heated until a coagulum is formed and this curd is filtered, washed, and separated. The watery residue (containing most of the salts) is discarded. The material recovered on the filter is the leaf protein. Figure 2 shows this process.

Atriplex triangularis produces leaves that are similar to spinach in appearance and nutritional composition. Estimated yields of 21.3 tons per hectare (fresh weight) have been reported using seawater for irrigation. (M.N. Islam)

Carlsson* observed that the expressed juice of some plants coagulated spontaneously at ambient temperatures. This reaction correlates with an undesirably high tannin and polyphenol content and can serve as screening technique to eliminate candidate plants.

Leaf protein can be used as an additive to enhance the protein content of many food products. In India, for example, leaf protein is cooked with sugar and corn flour to make a confection; in Mexico, it is used to make a fortified spaghetti. Other leaf protein facilities have been set up in villages in Bolivia, Ghana, Pakistan, and Sri Lanka.

In Sri Lanka, hand- and foot-powered presses are used to extract leaf protein from local plants. This leaf protein is used to fortify a local traditional dish, *kola kenda*, prepared from cooked rice and coconut. Children who received this fortified food were found to be significantly healthier than children from a nearby village who were

---

* R. Carlsson. 1983. Tropical plants for leaf protein concentrates. In: L. Telek and H. D. Graham (eds.) *Leaf Protein Concentrate*. AVI Publications, Westport, Connecticut, USA

Leaf protein can be extracted from the foliage of salt-tolerant plants. In Sri Lanka, hand-powered presses (top) and foot-powered presses (bottom) are used to extract leaf juice; this juice is then heated and leaf protein coagulates and is used to improve the nutrition of traditional foods. (A. Maddison)

not given leaf protein. After initial introduction in one village, the production and use of leaf protein spread to thirty villages.

In Ghana, a village cooperative was established to produce leaf protein for food use, silage from the fibrous residue, and alcohol from the residual liquid fraction. Leaf protein was sold at a price comparable with other protein-rich foods. Further economies (or

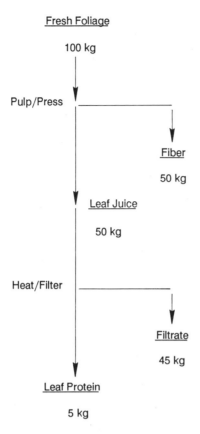

FIGURE 2 Leaf protein production. Freshly gathered leaves are pulped and pressed to yield juice and fiber fractions. The fiber can be used for ruminant feed. The juice is heated to coagulate the protein and this is filtered for use as a food supplement. Yield figures are typical of field results. SOURCE: Fellows, 1987.

profits) will be possible when income is obtained from the sale of the silage. Villagers participating in the cooperative derived as much as a fivefold increase in income.

Various salt-tolerant plants have been used for leaf protein production including *Kochia scoparia*, *Salsola kali*, *Beta maritima*, *Salicornia* spp., *Mesembryanthemum* spp., and *Atriplex* spp. (Carlsson, 1975). Some of the nutrients in leaf protein concentrate are shown in Table 5.

TABLE 5 Leaf Protein Composition.

Component	Per 100 g Dry Matter
True protein	50-60 g
Lipids	10-25 g
Beta Carotene	45-150 mg
Starch	2-5 g
Monosaccharides	1-2 g
B-vitamins	16-22 mg
Vitamin E	15 mg
Choline	220-260 mg
Iron	40-70 mg
Calcium	400-800 mg
Phosphorus	240-570 mg
Ash	5-10 g

SOURCE: Carlsson, 1988.

FRUITS

Ahmad* has described a technique developed in Pakistan and India to grow salt-sensitive fruits on saline land. This involves grafting a salt-sensitive shoot on a salt-tolerant rootstock. For example, shoots of *Ziziphus mauritiana* (salt sensitive, but yielding fleshy berries) have been grafted on the roots of *Z. nummularia* (salt tolerant, but yielding smaller berries) to allow fruit production on saline land. Similarly, shoots of *Manilkara zapota* (salt sensitive, but bearing large fruit) have been grafted on rootstocks of *M. hexandra* (salt tolerant, but bearing small fruit) to combine the desirable qualities of both. Pasternak (1987) reported that pear cultivars can tolerate irrigation water of 6.2 dS/m when grafted on a quince rootstock.

Salvadora persica and *S. oleoides* are small evergreen trees or shrubs. Both species yield edible fruits. Their seeds contain about 40 percent of an oil with a fatty acid composition (lauric, 20 percent; myristic, 55 percent; palmitic, 20 percent; oleic, 5 percent), which makes an excellent soap. The seed oil is inedible because of the presence of various substituted dibenzylureas. Both are multipurpose trees in India and Pakistan, providing fodder and wood as well as fruit. In India, *S. persica* occurs on saline soils and in coastal regions just above the high-water line. Before the introduction of canal

---

*Rafiq Ahmad, University of Karachi. Personal communication.

irrigation in Pakistan, *S. oleoides* occupied much of the worst salt-affected land.

There are about a dozen species of *Lycium* in the United States. Although most bear edible fruit, they are commonly cultivated as ornamentals. *L. fremontii* seems to have agronomic promise. It is a thorny shrub native to southern Arizona and the Gulf of California region in adjacent Mexico. It thrives on desert soils, upper beaches, and semisaline and alkaline flats both near the coast and on inland deserts.

The quandong (*Santalum acuminatum*) is widely distributed across Australia's arid inland. This small tree averaging about 4 m high, has bright red cherry-sized fruit with edible flesh and a stone with an edible kernel. The flesh is a good source of carbohydrate (19-23 percent). It was a staple of the aboriginal's diet and has been popular with other Australians in jam and pie. The kernel is roasted before being consumed and has a high oil (58 percent) and energy content. The quandong is reported to be highly resistant to drought, high temperatures, and salinity. An experimental orchard in southern Australia has been irrigated for seven years with water with a conductivity of 4.7 dS/m.

The seagrape (*Coccoloba uvifera*)* is readily established on sandy shores. When fully exposed on windswept seacoasts, the seagrape is dwarfed and bushy (to 2.5 m high) and forms dense colonies. Inland, it becomes a spreading, low-branched tree (to 15 m high). The wood makes excellent fuel and can also be used for furniture and cabinetwork. The fruits are popular in the Caribbean and are sold in local markets. The flowers yield abundant nectar and result in a fine honey.

TRADITIONAL CROPS

In Israel, a number of commercial crops are grown with underground brackish water. These include melons, tomatoes, lettuce, Chinese cabbage, and onions. A study on market tomatoes showed that fruits produced under saline conditions were smaller than the controls, but developed a better color and had a much better taste. However, their shelf life tended to be shorter. Taste testing of other crops grown in brackish water showed that in melons, the fresh fruits

---

*See also pp. 8-9 in *Firewood Crops: Shrub and Tree Species for Energy Production* (Volume 2). To order, see p. 135.

The quandong tree has bright red cherry-sized fruit with edible flesh and a stone with an edible kernel. Averaging about 4 m high, it is widely distributed across Australia's arid interior. It has been grown using saline water for irrigation. (M. Sedgley)

TABLE 6 Experimental Yields of Vegetables and Grains at the Ramat Negev Experimental Station.

Crop	System*	Yield (t/ha) at EC (irrigation water)					Species
		1.2	3.5-5.5	6-8	8-10	10-15	
Vegetables							
Asparagus	d	6.6	6.6	--	--	--	*A. officinalis*; 4-year-old plot.
Broccoli	d	23.4	21.8	--	19.0	14.3	*Brassica oleracea*
Beetroot	s	55.5	52.7	--	--	--	*Beta vulgaris*
Carrot	d	45.8	41.2	33.8	0	--	*Daucus carota*
Celery	s	155.0	171.0	--	--	--	*Apium graveolens*
Chinese cabbage	d	135.0	118.0	108.0	109.0	--	*Brassica pekinensis*
Chinese cabbage	d	58.0	58.0	55.0	65.0	--	*Brassica chinensis*
Kohlrabi	d	30.0	20.3	17.4	11.7	--	*Brassica caulorapa*
Lettuce	d	67.7	64.5	52.8	58.3	--	*Lactuca sativa*
Melon	d	27.0	24.0	24.0	22.0	--	*Cucumis melo*
Onion	d	50.1	28.4	4.1	0.4	--	*Allium cepa*
Onion	d	50.1	34.0	27.9	22.4	--	*A. cepa*; saline irrigation from 64th day after planting.
Tomato	d	86.5	72.9	--	62.7	53.0	*Lycopersicon esculentum*
Grains		(Yield of grain at 12% moisture)					
Maize	d	7.1	4.6	3.1	1.3	--	*Zea mays*
Maize	d	7.0	6.7	7.0	5.2	--	*Z. mays*; saline irrigation from 21st day after germination.
Sorghum	s	10.0	8.4	--	--	--	*Sorghum vulgare*
Wheat	s	6.8	6.7	--	--	--	*Triticum vulgare*

* d = drip irrigation, s = sprinkler irrigation.

SOURCE: Pasternak and De Malach, 1987.

were tastier than the controls. For lettuce, the salinity of the irrigation water had no discernible effect on the taste. Yields obtained in seventeen saline irrigation experiments are shown in Table 6.

Asparagus (*Asparagus officinalis*) is commonly considered a temperate crop, dormant in the winter with spears harvested in the spring, and summer fern growth terminated by cooler fall weather. In tropical areas it can be grown using the "mother fern" technique. After plants are established, the first two or three spears are allowed to grow to fern; thereafter, spears are harvested as they develop. Twice during the year old fern is replaced by new fern, but asparagus is produced year-round with annual yields exceeding those obtained in temperate climates.

In Tunisia, asparagus is grown near Zarzis, where the salinity the irrigation water is 6.5 g per liter. Yields (4-8 tons per hectare) are about the same as in areas irrigated with fresh water. It has also been grown experimentally in Israel's Negev desert.

In the United States, University of Delaware researchers found *A. officinalis* growing wild at the edge of a salt marsh. Using commercial asparagus varieties, they germinated thousands of seeds in fresh water and transferred the seedlings to salt water. Most died, but some grew well at salinities of 30 parts per thousand.

Asparagus is an excellent crop for developing countries because it is relatively labor intensive. Although several years are required before a marketable crop is obtained, production continues for 15-25 years. Water requirements for asparagus are somewhat greater than for cotton, and a light soil and careful management are required.

Rice (*Oryza sativa*) is a staple crop in many developing countries. It has been observed that coastal-grown rice generally gives lower yields than inland rice, presumably because of the effects of saline soil or salty ocean mists. Rice cells subjected to salt stress and then grown to maturity had progeny with improved salt tolerance—up to 1 percent salt.

Barley (*Hordeum vulgare*) is the most salt-tolerant cereal grain. At the University of Arizona, a special strain of barley yielded about 4,000 kg per hectare when irrigated with groundwater with half the salinity of seawater. At the University of California, specially selected strains of barley were grown on sand dunes with seawater and diluted seawater irrigation. Yields were 3,102 kg per hectare for fresh water, 2,390 kg per hectare for one-third seawater, 1,436 kg per hectare for two-thirds seawater, and 458 kg per hectare for full-strength seawater.

Wheat (*Triticum aestivum*) is an important source of human nutrition, and the improvement of salt tolerance in this crop deserves attention. Traditional cultivars from salt-affected areas may serve as sources for salt resistance in modern wheat varieties. There is a need to collect and evaluate cultivars from lands where salt stress has been exerting selection pressure over long periods. In India, researchers at the Central Soil Salinity Research Institute have collected and evaluated more than 400 indigenous cultivars from salt-affected regions of the Indian subcontinent.

In addition, many wild relatives of wheat show outstanding adaptation to saline environments. For example, tall wheatgrass (*Elytrigia [Agropyron] elongatum*) and *E. pontica* have been reported to survive salt concentrations higher than seawater. The salt tolerance of

Traditional wheat cultivars from salt-affected lands can be used to breed salt resistance in modern wheat varieties. (J.A. Aronson)

TABLE 7 Salt-Tolerant Plants for Honey Production.

Species	Honey Production (kg per colony per year)
Agave americana	41 (Mexico)
Cajanus cajan	--
Dalbergia sissoo	4-9 (India)
Eucalyptus camaldulensis	55-60 (Australia)
E. gomphocephala	--
E. paniculata	100 (Australia)
Gleditsia triacanthos	250* (Romania)
Lotus corniculatus	--
Parkinsonia aculeata	--
Pithecellobium dulce	--
Pongamia pinnata	--
Prosopis cineraria	--
P. pallida	120-363 (Hawaii)
Trifolium alexandrinum	165* (Bulgaria)

*Kg per season from one hectare covered with the plant.

SOURCE: Crane, 1985.

wheat may be enhanced through hybridization and selective transfer of gene complexes from these valuable resources.

Researchers at the Institute of Plant Science Research (Cambridge, England) have succeeded in crossing salt-tolerant sand couch (*Thinopyrum bessarabicum*) with wheat. Sand couch grows on the sand dunes of the Black Sea and can withstand salt concentrations that would be lethal for wheat. The sand couch/wheat hybrid can grow and set seed at salt levels of 1.1 percent.

A recent report by Rawson and coworkers (1988) suggests that absolute NaCl tolerance in wheat, barley, and triticale is not so much due to the greater ability to grow in the presence of NaCl, but to grow well *per se*. In many cases, productivity in NaCl can be estimated from the size of seedling leaves on the control plants.

Maas and coworkers (1983) have examined the effects of saline water on germination, growth, and seed production in maize (*Zea mays*). At germination, salinities of up to 10 dS/m can be tolerated, but dry matter production is decreased if the EC exceeds 1 dS/m during seedling growth. Increasing the salinity of the irrigation water to 9 dS/m at the tasseling and grain filling stages did not significantly reduce yields.

Some salt-tolerant plants are suitable for honey production, with the honey being used directly by the farmer or sold for added income. Although it would probably not be cost effective to establish salt-tolerant plants solely for honey production, it could be a valuable adjunct while plants are maturing for other uses. The black mangrove (*Avicennia germinans*), for example, has an intense summer flow of nectar heavily gathered by honeybees. Fourteen other tropical and subtropical plants that are valuable honey sources are listed in Table 7.

REFERENCES AND SELECTED READINGS

General

Downton, W. J. S. 1984. Salt tolerance of food crops: prospectives for improvements. *CRC Critical Reviews in Plant Sciences* 1(3):183-201.
Epstein, E. and D. W. Rains. 1987. Advances in salt tolerance. *Plant and Soil* 99:17-29.
Gallagher, J. L. 1985. Halophytic crops for cultivation at seawater salinity. *Plant and Soil* 89:323-336.
Gamborg, O. L., R. E. B. Ketchum and M. W. Nabors. 1986. Tissue culture and cell biotechnology for increased salt tolerance in crop plants. Pp. 81-92 in: R. Ahmad and A. San Pietro (eds.) *Prospects for Biosaline Research*. University of Karachi, Karachi, Pakistan.
Jain, S. C., R. K. Gupta, O. P. Sharma and V. K. Paradkar. 1985. Agronomic manipulation in saline sodic soils for economic biological yields. *Current Science* 54(9):422-425.
Maas, E. V. 1986. Crop tolerance to saline soil and water. Pp. 205-219 in: R. Ahmad and A. San Pietro (eds.) *Prospects for Biosaline Research*. University of Karachi, Karachi, Pakistan.
Mizrahi, Y. and D. Pasternak. 1985. Effect of salinity on quality of various agricultural crops. *Plant and Soil* 89:301-307.
O'Leary, J. W. 1985. Saltwater crops. *CHEMTECH* 15(9):562-566.
O'Leary, J. W. 1987. Halophytic food crops for arid lands. Pp. 1-4 in: *Strategies for Classification and Management of Native Vegetation for Food Production in Arid Areas*. Report RM-150, Forest Service, USDA, Ft. Collins, Colorado 80526, US.
Pasternak, D. 1987. Salt tolerance and crop production—a comprehensive approach. *Annual Review of Phytopathology* 25:271-291.
Pasternak, D. and Y. De Malach. 1987. Saline water irrigation in the Negev Desert. in: *Agriculture and Food Production in the Middle East*. Proceedings of a Conference on Agriculture and Food Production in the Middle East, Athens, Greece. January 21-26, 1987.
Somers, G. F. 1982. Food and economic plants: general review. Pp. 127-148 in: A. San Pietro (ed.) *Biosaline Research*. Plenum Press, New York, New York, US.

Grains and Oilseeds

Zostera marina

de Cock, A. W. A. M. 1980. Flowering, pollination and fruiting in *Zostera marina*. *Aquatic Botany* 9(3):210-220.

Felger, R. S. and C. P. McRoy. 1975. Seagrasses as potential food plants. Pp. 62-69 in: C. F. Somers (ed.) *Seedbearing Halophytes as Food Plants*. College of Marine Studies, University of Delaware, Newark, Delaware, US.

Thorhaug, A. 1986. Review of seagrass restoration efforts. *Ambio* 15(2):110-117.

Distichlis

Yensen, N. P., S. B. Yensen and C. W. Weber. 1985. A review of *Distichlis* spp. for production and nutritional values. Pp. 809-822 in: E. E. Whitehead, C. F. Hutchinson, B. N. Timmermann, and R. G. Varady (eds.) *Arid Lands Today and Tomorrow*, Westview Press, Boulder, Colorado, US.

Yensen, N. P. 1988. Plants for salty soil. *Arid Lands Newsletter* 27:3-10. University of Arizona, Tucson, Arizona, US.

Yensen, N. P. 1987. Development of a rare halophyte grain: prospects for reclamation of salt-ruined lands. *Journal of the Washington Academy of Sciences* 77(4):209-214.

Sporobolus airoides

Chadha, Y. R. (ed.). 1976. *Sporobolus. The Wealth of India* X:24-25. CSIR, New Delhi, India.

Doebley, J. F. 1984. "Seeds" of wild grasses: a major food of Southwestern Indians. *Economic Botany* 38:52-64.

Ezcurra, E., R. S. Felger, A. D. Russell and M. Equihua. 1988. Freshwater islands in a desert sand sea: the hydrology, flora, and phytogeography of the Gran Desierto oases of northwestern Mexico. *Desert Plants* 9(2):35-44,55-63.

Heizer, R. F. and A. B. Elsasser. 1980. *The Natural World of the California Indians*. University of California Press, Berkeley, California, US.

Quinoa

Atwell, W. A., B. M. Patrick, L. A. Johnson and R. W. Glass. 1983. Characterization of quinoa starch. *Cereal Chemistry* 60(1):9-11.

Risi, J. and N. W. Galwey. 1984. The *Chenopodium* grains of the Andes: Inca crops for modern agriculture. *Advances in Applied Biology* 10:145-216.

Kosteletzkya virginica

Gallagher, J. L. 1985. Halophytic crops for cultivation at seawater salinity. *Plant and Soil* 89:323-336.

Islam, M. N., C. A. Wilson and T. R. Watkins. 1982. Nutritional analysis of seashore mallow seed, *Kosteletzkya virginica*. *Journal of Agricultural and Food Chemistry* 30(6):1195-1198.

Acacias

Orr, T. M. and L. J. Hiddins. 1987. Contributions of Australian acacias to human nutrition. Pp. 112-115 in J. W. Turnbull (ed.) *Australian Acacias in Developing Countries*. ACIAR Proceedings no. 16. Canberra, Australia.

Brand, J. C., V. Cherikoff and A. S. Truswell. 1985. The nutritional composition of Australian Aboriginal bushfoods - 3, seeds and nuts. *Food Technology in Australia* 37:275-279.

Peterson, N. 1978. The traditional patterns of subsistence to 1975. Pp. 22-35 in: B. S. Hetzel and H. J. Frith (eds.) *The Nutrition of Aborigines in Relation to the Ecosystem of Central Australia*. CSIRO, Melbourne, Australia.

Terminalia catappa

Morton, J. F. 1985. Indian almond (*Terminalia catappa*), salt-tolerant, useful, tropical tree with "nut" worthy of improvement. *Economic Botany* 39:101-112.

Argan

Morton, J. F. and G. L. Voss. 1987. The argan tree (*Argania siderozylon*, Sapotaceae), a desert source of edible oil. *Economic Botany* 41:221-223.

Salicornia

Charnock, A. 1988. Plants with a taste for salt. *New Scientist* 120(1641):41-45.

Tubers and Foliage

Batis maritima

Glenn, E. P. and J. W. O'Leary. 1985. Productivity and irrigation requirements of halophytes grown with seawater in the Sonoran Desert. *Journal of Arid Environments* 9(1):81-91.

Sesuvium portulacastrum

Chadha, Y. R. (ed.). 1972. *Sesuvium*. The Wealth of India IX:304. CSIR, New Delhi, India.

Portulaca oleracea

Sen, D. N. and R. P. Bansal. 1979. Food plant resources of the Indian deserts. Pp. 357-370 in: J. R. Goodin and D. K. Northington (eds.) *Arid Plant Resources*. Texas Tech University, Lubbock, Texas, US.

Crithmum maritimum

Franke, W. 1982. Vitamin C in sea fennel (*Crithmum maritimum*), an edible wild plant. *Economic Botany* 36:163-165.

Okusanya, O. T. 1977. The effect of sea water and temperature on the germination behavior of *Crithmum maritimum*. *Physiologia Plantarum* 41(4):265-267.

Atriplex triangularis

Islam, M. N., R. R. Genuario and M. Pappas-Sirois. 1987. Nutritional and sensory evaluation of *Atriplex triangularis* leaves. *Food Chemistry* 25:279-284.

Khan, M. A. 1987. Salinity and density effects on demography of *Atriplex triangularis* Willd. *Pakistan Journal of Botany* 19(2):123-130.

Riehl, T. E. and I. A. Ungar. 1983. Growth, water potential, and ion accumulation in the inland halophyte *Atriplex triangularis* under saline field conditions. *Acta Oecologica, Oecologia Plantarum* 4:27-39.

Mesembryanthemum crystallinum

Sastri, B. N. (ed.). 1962. *Mesembryanthemum*. *The Wealth of India* VI:349. CSIR, New Delhi, India.

Suaeda maritima

Chadha, Y. R. (ed.). 1976. *Suaeda*. *The Wealth of India* X:70-71. CSIR, New Delhi, India.

Leaf Protein

Carlsson, R. 1988. *Leaf Nutrients for Human Consumption: A Global Overview* (Swedish). University of Lund, Lund, Sweden.

Carlsson, R. 1980. Quantity and quality of leaf protein concentrates from *Atriplex hortensis, Chenopodium quinoa* and *Amaranthus caudatus* grown in southern Sweden. *Acta Agriculturae Scandinavica* 30(4):418-426.

Carlsson, R. 1975. *Centrospermae Species and Other Species for Production of Leaf Protein*. Ph.D. thesis. University of Lund, Lund, Sweden.

Fellows, P. 1987. Village-scale leaf fractionation in Ghana. *Tropical Science* 27:77-84.

Martin, C. 1987. Leaf extract boosts nutritional value. *VITA News* (July):11-12.

Maddison, A. and G. Davys. 1987. Leaf protein - a simple technology to improve nutrition. *Appropriate Technology* 14(2):10-11.

Pirie, N. W. 1987. *Leaf Protein and its By-Products in Human and Animal Nutrition*. Cambridge University Press, New Rochelle, New York, US.

Singh, A. K. 1985. The yield of leaf protein from some weeds. *Acta Botanica Indica* 13(2):165-170.

Valenzuela, J. 1988. Protein for the young and needy. *South* 88:99.

Fruits

Salvadora

Gupta, R. K. and S. K. Saxena. 1968. Resource survey of *Salvadora oleoides* and *S. persica* for non-edible oil in western Rajasthan. *Tropical Ecology* 9:140-152.

Ezmirly, S. T. and J. C. Cheng. 1979. Saudi Arabian medicinal plants: *Salvadora persica. Planta Medica* 35(2):191-192.
Chadha, Y. R. (ed.). 1972. *Salvadora. Wealth of India* IX:193-195. CSIR, New Delhi, India

Lyciums

Felger, R. S. and M. B. Moser. 1984. *People of the Desert and Sea: Ethnobotany of the Seri Indians.* University of Arizona Press, Tucson, Arizona, US.
Greenhouse, R. 1979. *The Iron and Calcium Content of Some Traditional Pima Foods and the Effects of Preparation Methods.* (Thesis) Arizona State University, Tempe, Arizona, US.

Santalum acuminatum

Jones, G. P., D. J. Tucker, D. E. Rivett and M. Sedgley. 1985. The nutritional potential of the quandong (*Santalum acuminatum*) kernel. *Journal of Plant Foods* 6(4):239-246.
Possingham, J. 1986. Selection for a better quandong. *Australian Horticulture* 84(2):55-59.
Sedgley, M. 1982. Preliminary assessment of an orchard of quandong seedling trees. *Journal of the Australian Institute of Agricultural Science* 48:52-56.

Traditional Crops

Asparagus

Nichols, M. A. 1986. Asparagus coming into its own as a high-value field crop. *Agribusiness Worldwide* 6(8):15-18.
Robb, A. 1984. Asparagus production using mother fern. *Asparagus Research Newsletter* (New Zealand) 2(1):24.

Rice

Akbar, M. 1986. Breeding for salinity tolerance in rice. Pp. 37-55 in: R. Ahmad and A. San Pietro (eds.) *Prospects for Biosaline Research.* University of Karachi, Karachi, Pakistan.
Dubey, R. S. and M. Rani. 1989. Influence of NaCl salinity on growth and metabolic status of protein and amino acids in rice seedlings. *Journal of Agronomy and Crop Science.* 162(2):97-106.
Ponnamperuma, F. N. 1984. Role of cultivar tolerance in increasing rice production on saline lands. in: R. C. Staples & G. H. Toenniessen (eds.) *Salt Tolerance in Plants.* John Wiley, New York, New York, US.
Wong, C.-K., S.-C. Woo and S.-W. Ko. 1986. Production of rice plantlets on NaCl-stressed medium and evaluation of their progenies. *Botanical Bulletin Academia Sinica* 27:11-23.

Barley

Anonymous. 1982. New variety yields 1.2 tonnes/ha when irrigated from the ocean. *International Agricultural Development* 2(3):29.

Iyengar, E. R. R., J. Chikara and P. M. Sutaria. 1984. Relative salinity tolerance of barley varieties under semi-arid climate. *Transactions of Indian Society of Desert Technology* 9(1):27-33.

Norlyn, J. D. and E. Epstein. 1982. Barley production: irrigation with seawater on coastal soil. Pp. 525-529 in: A. San Pietro (ed.) *Biosaline Research*. Plenum Press, New York, New York, US.

Wheat

Dvorak, J., K. Rose and S. Mendlinger. 1985. Transfer of salt tolerance from *Elytrigia pontica* to wheat by the addition of an incomplete *Elytrigia* genome. *Crop Science* 25:306-309.

Forster, B. 1988. Wheat can take on more than a pinch of salt. *New Scientist* 120(1641):43.

Gorham, J., E. McDonnell and R. G. Wyn Jones. 1984. Salt tolerance in the Triticeae: *Leymus sabulosus*. *Journal of Experimental Botany* 35:1200-1209.

Gulick, P. and J. Dvorak. 1987. Gene induction and repression by salt treatment in the roots of the salinity-sensitive Chinese Spring wheat and the salinity-tolerant Chinese Spring x *Elytrigia elongata* amphiploid. *Proceedings of the National Academy of Sciences* 84:99-103.

Maas, E. V. and J. A. Poss. 1989. Salt sensitivity of wheat at various growth stages. *Irrigation Science* 10:29-40.

Rana, R. S. 1986. Genetic diversity for salt-stress resistance of wheat in India. *Rachis* 5(1):32-37.

Rana, R. S. 1986. Evaluation and utilisation of traditionally grown cereal cultivars of salt affected areas of India. *Indian Journal of Genetics* 46:121-135.

Rawson, H. M., R. A. Richards and R. Munns. 1988. An examination of selection criteria for salt tolerance in wheat, barley and triticale genotypes. *Australian Journal of Agricultural Research* 39:759-792.

Sajjad, M. S. 1986. Evaluation of wheat germplasm for salt tolerance. *Rachis* 5(1):28-31.

Maize

Ahmad, R., S. Ismail and D. Khan. 1986. Use of highly saline water for irrigation at sandy soils. Pp. 389-413 in: R. Ahmad and A. San Pietro (eds.) *Prospects for Biosaline Research*. University of Karachi, Karachi, Pakistan.

Maas, E. V., G. J. Hoffman, G. D. Chaba, J. A. Poss and M. C. Shannon. 1983. Salt sensitivity of corn at various growth stages. *Irrigation Science* 4:45-57.

Pasternak, D., Y. De Malach and I. Borovic. 1985. Irrigation with brackish water under desert conditions. II. Physiological and yield response of maize (*Zea mays*) to continuous irrigation with brackish water and to alternating brackish-fresh-brackish water irrigation. *Agricultural Water Management* 10:47-60.

Pessarakli, M., J. T. Huber and T. C. Tucker. 1989. Dry matter yields, nitrogen absorbtion, and water uptake by sweet corn under salt stress. *Journal of Plant Nutrition* 12(3):279-290.

Totawat, K. L. and A. K. Mehta. 1985. Salt tolerance of maize and sorghum genotypes. *Annals of Arid Zone Research* 24(3):229-236.

Tomato

Mizrahi, Y. 1982. Effect of salinity on tomato fruit ripening. *Plant Physiology* 69:966-970.

Jones, R. A. 1987. Genetic advances in salt tolerance. Pp. 125-138 in: D. J. Nevins & R. A. Jones (eds.) *Tomato Biotechnology.* Alan R. Liss, Inc., New York, New York, US.

Onion

Miyamoto, S. 1989. Salt effects on germination, emergence, and seedling mortality of onion. *Agronomy Journal* 81(2):202-207.

Honey

Crane, E. 1985. Bees and honey in the exploitation of arid land resources. Pp. 163-175 in: G. E. Wickens, J. R. Goodin and D. V. Field (eds.) *Plants for Arid Lands.* George Allen & Unwin, London, UK.

Morton, J. F. 1964. Honeybee plants of South Florida. *Proceedings of the Florida State Horticultural Society* 77:415-436.

RESEARCH CONTACTS

General

Rafiq Ahmad, Department of Botany, University of Karachi, Karachi 32, Pakistan.

James Aronson, 12 rue Vanneau, 34000 Montpellier, France.

Akissa Bahri, Centre de Recherches du Genie Rural, BP No. 10, Ariana 2080, Tunisia.

John L. Gallagher, College of Marine Studies, University of Delaware, Lewes, DE 19958, US.

Oluf L. Gamborg, Tissue Culture for Crops Project, Colorado State University, Ft. Collins, CO 80523, US.

E. R. R. Iyengar, Central Salt and Marine Chemicals Research Institute, Bhavnagar 364 002, India.

T. N. Khoshoo, Department of Environment, Bikaner House, Shahjahan Road, New Delhi 110 011, India.

Gwyn Jones, Human Nutrition Section, Deakin University, Victoria 3217, Australia.

S. Miyamoto, Texas Agricultural Experiment Station, 1380 A&M Circle, El Paso, TX 79927, US.

Yosef Mizrahi, Boyko Institute for Research, Ben Gurion University, PO Box 1025, Beer-Sheva 84110, Israel.

Gary P. Nabhan, Office of Arid Lands Studies, University of Arizona, Tucson, AZ 85719, US.

Dov Pasternak, Institute for Desert Research, Ben Gurion University, Sede Boger 84990, Israel.

James D. Rhoades, USDA Salinity Research Laboratory, Riverside, CA 92501, US.

M. C. Shannon, USDA Salinity Research Laboratory, Riverside, CA 92501, US.
G. E. Wickens, Royal Botanic Gardens, Kew, Richmond, Surrey TW9 3AE, UK.
Xie Cheng-Tao, Institute of Soil and Fertilizers, 30 Baishiqiao Road, Beijing 100081, People's Republic of China.

Grains and Oilseeds

Zostera marina

Richard S. Felger, Office of Arid Lands Studies, University of Arizona, Tucson, AZ 85719, US.

Distichlis

N. Yensen, NyPa, Inc., 727 North Ninth Avenue, Tucson, AZ 85705 US.

Quinoa

Rolf Carlsson, Institute of Plant Physiology, University of Lund, Box 7007, S-220 07 Lund, Sweden.
Instituto Interamericano de Ciencias Agricolas OEA, Andean Zone, Box 478, Lima, Peru.
John McCamant, Sierra Blanca Associates, 2560 South Jackson, Denver, CO 80210, US.
Ministerio de Asuntos Campesinos y Agropecuarios, Biblioteca Nacional Agropecuria, La Paz, Bolivia.
E. J. Weber, Agriculture, Food and Nutrition Division, IDRC Regional Office, Apartado Aereo 53016, Bogota, Colombia.

Pennisetum typhoides

E. R. R. Iyengar, Central Salt and Marine Chemicals Research Institute, Bhavnagar 364 002, India.

Kosteletzkya virginica

J. L. Gallagher, College of Marine Studies, University of Delaware, Lewes, DE 19958, US.
M. N. Islam, Department of Food Science, University of Delaware, Newark, DE 19716, US.

Acacia

Janette C. Brand, University of Sydney, Sydney, NSW 2006, Australia.

Tecticornia

Paul G. Wilson, Western Australian Herbarium, PO Box 104, Como, WA 6152, Australia.

Terminalia catappa

Julia F. Morton, Director, Morton Collectanea, University of Miami, Coral Gables, FL 33124, US.

Argan

Julia F. Morton, Director, Morton Collectanea, University of Miami, Coral Gables, FL 33124, US.

Salicornia

James O'Leary, University of Arizona, Tucson, AZ 85719, US
Carl Hodges, Environmental Research Laboratory, Tucson International Airport, Tucson, AZ 85706

Leaf Protein

Walter Bray, 13-15 Frognal, London NW3 6AP, UK.
Rolf Carlsson, Institute of Plant Physiology, University of Lund, Box 7007, S-220 07 Lund, Sweden.
Peter Fellows, Oxford Polytechnic, Gipsy Lane, Oxford OX3 OPB, UK
Shoaib Ismail, Department of Botany, University of Karachi, Karachi 32, Pakistan.
Carol Martin, Find Your Feet, 345 West 21st Street, Suite 3D, New York, NY 10011, US.
A. K. Singh, S 4/50 D4, Tajpur, Orderly Bazar, Varanasi, 221 002, India.

Fruits

Quandong

Margaret Sedgley, Waite Agricultural Research Institute, University of Adelaide, Glen Osmond, SA 5064, Australia.

Lycium

Richard S. Felger, Office of Arid Lands Studies, University of Arizona, Tucson, AZ 85719, US.

Coccoloba uvifera

Centro Agronomico Tropical de Investigacion y Ensenaza, Turrialba, Costa Rica.
Institute of Tropical Forestry, PO Box AQ, Rio Piedras, Puerto Rico 00928, US.
Instituto Forestal Latino-Americano, Apartado 36, Merida, Venezuela.

Traditional Crops

Asparagus

Yoel De Malach, Ramat Negev Regional Experimental Station, Doar Na Halutza 85515 Israel.
M. A. Nichols, Department of Horticulture and Plant Health, Massey University, Palmerston North, New Zealand.

Rice

I. U. Ahmed, Department of Soil Science, University of Dhaka, Dhaka 2, Bangladesh.
M. Akbar, International Rice Research Institute, PO Box 933, Manila, Philippines.
F. N. Ponnamperuma, International Rice Research Institute, PO Box 933, Manila, Philippines.
R. S. Rana, Genetics Research Center, Central Soil Salinity Research Institute, Karnal 132001, India.
C.-K. Wong, Institute of Botany, Academia Sinica, Nankang, Taipei, Taiwan.

Barley

E. Epstein, Department of Land, Air and Water Resources, University of California, Davis, CA 95616, US.
R. T. Ramage, College of Agriculture, University of Arizona, Tucson, AZ 85721, US.
E. R. R. Iyengar, Central Salt and Marine Chemicals Research Institute, Bhavnagar 364 002, India.
H. M. Rawson, Division of Plant Industry, CSIRO, PO Box 1600, Canberra, ACT 2601, Australia

Wheat

E. Epstein, Department of Land, Air and Water Resources, University of California, Davis 95616, CA, US.
S. Jana, Department of Crop Science and Plant Ecology, University of Saskatchewan, Saskatoon SN7 0W0, Canada.
R. Munns, Division of Plant Industry, CSIRO, PO Box 1600, Canberra, ACT 2601, Australia
R. S. Rana, Genetics Research Center, Central Soil Salinity Research Institute, Karnal 132001, India.
M. Siddique Sajjad, Nuclear Institute for Agriculture and Biology, PO Box 128, Faisalabad, Pakistan.
J. P. Srivastava, Cereal Improvement Program, International Center for Agricultural Research in Dry Areas, Aleppo, Syria.
R. G. Wyn Jones, Department of Biochemistry and Soil Science, University College of North Wales, Bangor LL57 2UW, Wales, UK.

Maize

D. Khan, Shoaib Ismail, Department of Botany, University of Karachi, Karachi 32, Pakistan.

Yoel De Malach, Ramat Negev Regional Experimental Station, Doar Na Halutza 85515 Israel.
K. L. Totawat, Department of Soil Science, Rajasthan College of Agriculture, Udaipur 313 001, India.

Tomato

Yoel De Malach, Ramat Negev Regional Experimental Station, Doar Na Halutza 85515 Israel.
Richard A. Jones, University of California, Davis, CA 95616, US.

Honey

Eva Crane, International Bee Research Association, Hill House, Gerrards Cross, Bucks SL9 ONR, UK.

2
Fuel

INTRODUCTION

More than a billion people in developing countries rely on wood for cooking and heating. In most developing areas, the rate of deforestation for fuelwood and for agricultural expansion far exceeds the rate of reforestation. People spend increasing amounts of time and money to acquire fuel. Substitute energy sources such as kerosene or electricity are either unavailable or are too expensive.

The need for agricultural land to feed growing populations makes it unlikely that high-quality land will be used for planting trees. There are several options for increasing the production of fuelwood—for example, improving cultural practices in existing forests, growing trees with other crops (agroforestry), and utilizing marginal lands.

Fuelwood and building materials can be produced from salt-tolerant trees and shrubs using land and water unsuitable for conventional crops. Fuel crop plantations established on saline soils or irrigated with saline water would allow better land and fresh water to be reserved for food or forage production. Moreover, some salt-tolerant *Prosopis*, *Eucalyptus*, and *Casuarina* can survive prolonged exposure to 40-45°C—temperatures that few food crops can withstand.

With careful planning, trees can help rehabilitate degraded lands by stabilizing the ecosystem and by providing niches and protection for other plants and animals. Criteria for selecting plant species for use as fuelwood in saline environments include:

- Rate of Growth and Regrowth—Although many species may survive in saline habitats, their growth is often too slow to provide any significant production. The ability to coppice is of great practical importance. Combustible litter and branches shed from some species is an advantage. High-density wood is preferred, but there is generally a negative correlation between density and growth rate. Species should be chosen that are easy to handle, cut, and split. The wood should burn evenly and slowly without sparks or noxious smoke.
- Establishment—In saline environments, establishment may be difficult. There may only be a brief period suitable for planting. Special preparation such as mulching, furrowing, or ridging may be required to facilitate early growth. Some halophytes can tolerate harsher conditions later in their growth than at germination.
- Adaptability—Some species require specialized habitats or microclimates and will not survive in all elements of the landscape or across an entire climatic zone. Plants with significant plasticity in climatic and site tolerance have greater potential for success.
- Diverse Use—Salt-tolerant trees and shrubs can serve other purposes. They can reduce wind erosion, protect row crops, provide shade or forage for livestock, and serve as a first step in land restoration. Spiny salt-tolerant shrubs can be planted as living fences. Trees can also serve to control salinity through their ability to use more water than crops or pasture on an annual basis, and to draw it from deeper in the soil profile. Candidate species that provide such benefits in addition to fuel production would be advantageous.

Since it is unlikely that any species will meet all these requirements, compromise is necessary. Although selection is usually based on performance in a similar environment, some species "travel" poorly, some show extreme variation in regard to source (provenance), and some perform remarkably well far outside their native climate.

In Australia, a consortium of business and academic groups* is developing a multitiered approach to provide salt-tolerant trees for

---

*Tree Tech Australia, PO Box 252, Applecross, WA 6153, Australia.

use as fuel and for pulp. The project will screen Australian trees for growth rates and salt and drought tolerance. In addition, root fungi, which help plants to obtain nutrients from the soil, will be screened for salt tolerance and their influence on tree growth. Trees with superior growth on saline soils will be tissue cultured and inoculated with salt-tolerant root fungi. These cloned trees will then be tested for field performance. After the field trials, useful plant material will be made commercially available for use in saline environments in Australia and other countries.

In the United States* *Eucalyptus* and *Casuarina* trees have been tested for over four years in demonstration plantations in California to reduce agricultural drainage water and lower water tables on saline sites. Superior trees have been cloned to produce seed and biomass for economic exploitation. Another Australian tree, *Acacia melanoxylon*, is also being evaluated in this project.

FUELWOOD TREES AND SHRUBS**

Some of the species that are promising for fuel production in saline environments are found in the genera *Prosopis*, *Eucalyptus*, *Casuarina*, *Rhizophora*, *Melaleuca*, *Tamarix*, and *Acacia*.

Prosopis

Shrubs and trees of the genus *Prosopis* are found throughout arid and semiarid areas of the tropics. Since they fix nitrogen, they improve the soil and so supply part of their own nutrients. In tests by Felker and coworkers (1981), *P. articulata*, *P. pallida*, and *P. tamarugo* all grew and fixed nitrogen when irrigated with water containing 3.6 percent salt.

In more recent tests by Rhodes and Felker (1988), *Prosopis* seeds from widely divergent saline areas of Africa, Argentina, Chile, Mexico, and the United States were germinated and grown in sand

---

*Roy M. Sachs, Department of Environmental Horticulture, University of California, Davis, CA 95616.

**Additional information on many of these fuelwood species can be found in *Firewood Crops: Shrub and Tree Species for Energy Production*, *Casuarinas: Nitrogen-Fixing Trees for Adverse Sites*, *Mangium and Other Fast-Growing Acacias for the Humid Tropics*, and *Tropical Legumes: Resources for the Future*. To order, see p. 135.

culture at NaCl concentrations up to 3.3 percent. Six of the species tested had seedlings that grew in 3.3 percent NaCl. *P. juliflora*, from West Africa, seemed to have the best potential for rapid growth at high salinity. Other *Prosopis* surviving at 3.3 percent NaCl were *P. chilensis*, *P. articulata*, *P. alba*, *P. nigra/flexuosa*, and *P. alba/nigra*. *P. tamarugo*, identified earlier as having exceptional salt tolerance, died from stem fungal disease before salt was introduced. *P. pubescens* seedlings succumbed at 1.2 percent NaCl, possibly from fungal disease as well.

P. juliflora has few soil and water constraints. It can be grown in either dry or waterlogged saline areas, and on degraded soils with low fertility. A thorny, deciduous, large-crowned, deep-rooted tree, *P. juliflora* may grow to 10 m or more, depending on the variety and site. It is native to Central America and northern South America, but it has been widely propagated in Africa and Asia, particularly in India.

In India, *P. juliflora* has spread throughout the state of Tamil Nadu where it is used for fuel by many of the rural poor, and its availability is credited with a reduction of cutting in natural forests. In one district, where substantial saline patches occur, farmers use *P. juliflora* as a fallow species for four years. The trees are harvested for fuelwood or, in many cases, converted to charcoal. The land can then be used for food crops for at least two years, after which trees are replanted.

In Pakistan, more than 300 hectares of *P. juliflora* have been successfully established in sandy plains and dunes along the seacoast. Nursery-grown seedlings were irrigated with underground saline water for two years. After this, irrigation was discontinued, but the plants continued to grow well, using their extensive root systems to absorb rainwater and dew. Simultaneous plantings of *P. juliflora* in non-sandy strata with poorer percolation did not fare as well, apparently because of salt buildup in the root system. The wood produced in the sandy environment had a high heat content and low ash, indicating its suitability as fuelwood.

Many other species of *Prosopis* yield good fuelwood as well. *P. chilensis* has been planted extensively in arid areas and *P. alba* has been used for reforesting dry saline areas. *P. ruscifolia* and *P. pallida* also have potential for use on saline soils and *P. cineraria* tolerates soils with a pH of over 9.

About 9,000 hectares of *Prosopis* have been planted in the

In Pakistan, more than 300 hectares of *Prosopis juliflora* have been established in sandy plains and dunes along the coast. (R. Ahmad)

Beehive kilns are used to produce charcoal in coastal Pakistan. (R. Ahmad)

Bhavnagar area of India. Half is used by villagers for fuelwood and half belongs to the forestry department.

Eucalyptus

Of the more than 500 species of *Eucalyptus*, relatively few are salt tolerant. Among those that are salt tolerant, there is a broad range of adverse environments where they occur. For example, a recently described and appropriately named species, *E. halophila*, occurs on the edges of salt lakes in Australia. *E. angulosa* grows in white coastal sand in Western and South Australia. It is used as a windbreak in coastal areas and may be grown where salt spray is a problem. *E. torquata* occurs in South Australia, often on shallow rocky soils and in association with *Atriplex* species.

E. camaldulensis grows widely in arid areas, usually along permanent or seasonal inland streams. An Australian native, it is now planted in many Mediterranean countries and is used for fuelwood, charcoal, poles, and for paper and particleboard manufacture. It is adapted to tropical and temperate climates and will grow well on poor soils and in areas where there are prolonged dry seasons (provided its roots can reach groundwater) or where periodic waterlogging occurs. It is not suitable for planting in humid tropical lowlands, nor in coastal areas where it would be exposed to windblown salt. Of its numerous provenances, a few have been shown to be highly salt-tolerant.

E. occidentalis is drought resistant and tolerates high temperatures, salinity, and waterlogging. In Western Australia, it has been found in clayey soils adjacent to salt lakes. *E. sargentii* is also native to Western Australia, where it is frequently found in areas where salt appears on the soil surface. It is reported to be one of the hardiest species and one of the last to die in areas of increasing salinity.

Of several Australian eucalyptus species tested in Israel, the highest growth rate and resistance to salinity (~30 dS/m) were shown by *E. occidentalis* and *E. sargentii*; at lower salinity levels (20-30 dS/m), *E. spathulata*, *E. kondininensis*, and *E. loxophleba* also exhibited rapid growth.

Eucalyptus species reported by Blake (1981) to survive salt concentrations of ~1.8 percent are *E. calophylla*, *E. erythrocorys*, *E. incrassata*, *E. largiflorens*, *E. neglecta*, and *E. tereticornis*. Other species that have been reported to grow well in saline environments are listed in Table 8.

Eucalyptus sargentii is native to Western Australia. It is frequently found where salt appears on the soil surface and is one of the hardiest species in areas of increasing salinity. (G. Shay)

Casuarina

Casuarina equisetifolia is a fast-growing evergreen tree, 15-30 m tall, with a long straight trunk, 60-120 cm in diameter. It is native to southern Asia, Malaysia, coastal Queensland, Australia, and other Pacific Islands. It is an important fuelwood species in India and serves to stabilize coastal dunes in China. It has been successfully introduced to coastal East and West Africa and to many areas of the

Mangrove forests survive waterlogging, salinity, and strong coastal winds. They help protect shorelines and serve as nurseries for many fish species. (NOAA)

Caribbean. It can grow on loose seashore sand within a few meters of high tide.

Its success as an introduced species is due to its ability to grow on nutrient-poor soils and to tolerate windblown salt, high alkalinity levels (pH 9.0-9.5), and moderate groundwater salinity.

In a study of the effect of salinity on growth and nitrogen fixation in *C. equisetifolia*, it was found that increasing the NaCl level to 200 mM (about 1.2 percent) had little effect on nitrogen fixation. At intermediate levels of salinity (50-100 mM NaCl), nitrogen fixation and growth were greater than for the control.

Not all species of *Casuarina* are salt tolerant and there is significant variation among those that are. *C. cristata*, *C. glauca*, and *C. obesa* are all reported to be more salt tolerant than *C. equisetifolia* and more suitable for heavier clay soils and waterlogged conditions. In recent testing for performance in saline-waterlogged conditions, *C. obesa* grew better than *Eucalyptus camaldulensis* and five other Eucalyptus species (van der Moezel et al., 1988). *C. obesa* is noted for its ability to grow in warm subhumid and semiarid zones. It produces good fuelwood and is useful in shelterbelts.

TABLE 8 Salt-Tolerant *Eucalyptus* Species.

Eucalyptus Species	Other Site Characteristics
E. astringens	Dry
E. brockwayi	Dry
E. calycogona	Dry
E. campaspe	Dry
E. concinna	Dry
E. diptera	Dry, Coastal
E. flocktoniae	Dry
E. forrestiana	Dry, Coastal
E. gracilis	Dry, Clay
E. griffithsii	Dry
E. lehmannii	Dry, Coastal
E. (foecunda) leptophylla	Dry
E. lesouefi	Dry
E. longicornis	Dry
E. merrickiae	Dry
E. ovularis	Dry
E. platycorys	Dry
E. platypus	Dry
E. salmonophloia	Dry
E. woodwardii	Dry

SOURCE: Chippendale, 1973.

Rhizophora

Mangrove forests grow on 45 million hectares of tropical coastal and estuarine areas. They are tolerant of waterlogging, high salinity and humidity, and strong coastal winds. Although seawater is tolerated, most species grow best at lower salinity levels, particularly where there is freshwater seepage to moderate seawater salinity. Studies on the mangrove *Avicennia marina*, indicate that growth is poor in fresh water; maximum biomass production occurs at salinity levels of 25-50 percent of seawater.

Rhizophora species range from small shrubs to tall trees. While *R. mangle* and *R. mucronata* are usually about 20-25 m tall, *R. apiculata* can grow to heights of 60 m.

The principal use for most *Rhizophora* species is for fuelwood and charcoal. Most species also produce a strong, attractive timber, notably durable in water. Mangroves have the added value of reducing typhoon damage, binding and building sand and soil, serving as spawning and nursery grounds for many species of fish and shellfish, and as nesting and feeding sites for seabirds. Mangroves serve as a

In addition to fuel use, mangroves are cut for boat construction (top) and for conversion to paper pulp (bottom). (WWF Photolibrary/Xavier Lecoultre)

special link between the land and sea; inorganic nutrients from the land become organic nutrients and are passed on to the sea.

R. mangle has been planted for coastal protection in Florida and Hawaii. *R. mucronata* is used for replanting cleared areas in Malaysia. Mangrove swamps have been managed for fuelwood in Malaysia for more than 80 years with harvest on a 30-year cycle. In Indonesia, the rotation is 20 years for firewood and 35 years for charcoal. In Thailand, a 30-year rotation is practiced for producing poles, firewood, and charcoal.

The black mangrove, *Avicennia germinans*, of the New World tropics and subtropics, as well as the Old World species *A. marina* and *A. officinalis*, inhabiting salt marshes, tidal swamps, and muddy coasts, provide fuel, charcoal, and wood for boats, furniture, posts, pilings, and utensils.

Mangroves are generally slow growing and cannot tolerate indiscriminate lopping. Although some species can be established by direct seeding, if strip-felling rather than clear-cutting is used for harvest, natural regeneration will occur. The Mangrove Research Center* in the Philippines has a mangrove nursery and a working group on the silviculture of mangroves.

Melaleuca

Melaleuca quinquenervia and *M. viridiflora* are often found together occupying slightly higher ground next to mangrove swamps. *M. quinquenervia* is deep rooted and can grow on nutrient-poor coastal soils. It can grow near the beach and survives windblown salt. Although it grows best in fresh water, it can tolerate saline groundwater. It is an excellent fuelwood and regenerates readily after coppicing. It seeds profusely and can become a nuisance in areas where occasional fires create a suitable seedbed.

M. styphelioides is a fast-growing tree, 6-18 m tall, found in swampy coastal sites in eastern Australia. It is more salt tolerant than *M. quinquenervia* and tolerates a wide variety of conditions including sandy, wet, saline, and heavy clay soils and some coastal exposure.

---

*Mangrove Research Center, Forest Research Institute, Laguna 3720, Philippines.

Six species of *Melaleuca* from an area of salt lakes in Western Australia were examined for their relative salt tolerance in greenhouse tests. Growth and survival at salinity levels up to 7.2 dS/m were tracked over 15 weeks. *M. cymbifolia* had the highest survival rate and *M. thyoides* the best growth in these tests. *M. thyoides*, a large shrub, also has outstanding tolerance to waterlogging.

M. bracteata, M. calycina, M. cardiophylla, M. glomerata, M. nervosa, M. pauperiflora, and *M. subtrigona* also occur on the margins of salt lakes in the interior of Australia.

Tamarix

Tamarisks are hardy shrubs or trees of the desert and seashore. There are more than 50 species of tamarisk and most tolerate salty soils, poor-quality water, drought, and high temperatures. Several types can be used to afforest sand dunes and saline wastelands. They have been used as windbreaks in desert areas and the mature trees can be used for lumber and fuelwood.

One disadvantage of tamarisks is the high salt content of their litter and the salt drip from their leaves. Vegetation surrounding these trees is killed and, where they are planted as a windbreak for agriculture, an open space must be allowed between the trees and the crop to prevent yield reduction. Leaves and twigs will not burn because of their high salt content. These drawbacks must be weighed against their useful characteristics when considering their introduction.

Tamarix aphylla is a heavily branched tree, 8-12 m tall at maturity. It has a deep and extensive root system and, like other *Tamarix* species, it excretes salt. Salty "tears" drip from the glands in its leaves at night, so that the soil under the tree is covered with salt. Field tests in Israel showed that *T. aphylla, T. chinensis,* and *T. nilotica* could all be grown with seawater irrigation.

T. stricta is a tree from the Middle East, closely resembling *T. aphylla,* but *T. stricta* has straighter stems, a denser canopy, and faster growth. *T. articulata* and *T. gallica* are reported to grow well on moderately salty sites in Western Australia. Both can be readily propagated from cuttings.

In a study of biomass production using tamarisks irrigated with saline water, Garrett (1979) found that *T. aphylla* had a higher growth rate than *T. africana* or *T. hispida.* He also projected *T.*

Field tests in Israel have shown that *Tamarix chinensis* can be grown with seawater irrigation. (G. Shay)

aphylla yields of up to 14 dry tons per acre when irrigated with 0.06-3.5 percent saline water.

Acacia

More than 800 of the known species of *Acacia* are native to Australia and many have potential for establishment on salt-affected sites. Species such as *A. longifolia*, *A. saligna*, and *A. sophorae* have been used to stabilize dunes in Israel and North Africa. In tropical

Australia, *A. oraria* grows close to the sea and *A. crassicarpa*, in association with *Casuarina equisetifolia*, tolerates salt-laden winds on frontal sand dunes.

Some *Acacia* species tolerate high levels of groundwater salinity. *A. stenophylla* is widely planted on salt-affected sites and *A. redolens, A. ampliceps, A. xiphophylla*, and *A. translucens* all grow in highly saline areas. Other species with good salt resistance include *A. floribunda, A. pendula, A. pycnantha, A. retinodes*, and *A. cyclops*.

A. auriculiformis is suitable for coastal sandy sites subjected to windblown salt and areas with acid or alkaline conditions. In northern Australia, it grows on sand dunes with a soil pH of 9.0. In laboratory tests, it has tolerated highly acid conditions. This nitrogen-fixing species also grows well in seasonally waterlogged areas. It has the disadvantage of brittle branches, which may break in ordinary winds.

Other Species

In India, twenty species of trees and shrubs were planted in a trial using saline water (EC = 4.0-6.1 dS/m) for irrigation. Of these, nine species were growing well after 18 months. The trees included *Acacia nilotica, Albizzia lebbek, Cassia siamea, Pongamia pinnata, Prosopis juliflora, Syzygium cumini*, and *Terminalia arjuna*; shrubs were *Adhatoda vasica* and *Cassia auriculata*. On the basis of costs for establishing and maintaining these plants, and the selling price for firewood, it was estimated that the required investment would be recovered in five years.

Pongamia pinnata, known as karanja, is found along the banks of streams and rivers and in beach and tidal forests in India. In West Bengal, a rotation of 30 years is used in *Pongamia* fuelwood plantations. Pongam oil, 27-39 percent of the seed, is used for leather treatment, soap making, lubrication, and medicinal purposes. An active component in the oil, karanjin, is reported to have insecticidal and antibacterial properties.

Butea monosperma is a medium-sized (3-4 m) deciduous tree that grows in waterlogged and saline soils in tropical Asia. Its profuse spring canopy of scarlet flowers earn it the common name "flame of the forest." Its seeds and seed oil have anthelmintic properties.

The Manila tamarind (*Pithecellobium dulce*) is a hardy evergreen tree that grows to 18 m in the Indian plains and tropical Americas. A legume, it grows in poor and sandy soils and survives in coastal

The Manila tamarind can grow in poor and sandy soils, and survives in coastal areas even with its roots in salty water. (G. Shay)

areas even with its roots in salty water. It is also drought resistant and reproduces readily from seeds and cuttings. In addition to its fuel use, the fleshy pulp of its pods is consumed as a fruit and its leaves and pods are used as fodder for cattle, sheep, and goats.

Information on more than 1,500 species of ground cover, vines, grasses, herbs, shrubs, and trees that tolerate seashore conditions has been assembled by Menninger (1964). Most of these are categorized as to their ability to grow (1) right on the shore, (2) with some protection, or (3) well back from the beach. Although there are no indications of other desirable characteristics for use as fuelwood, some of the trees suggested for planting where there is direct exposure to salt spray and sand (category 1) are listed in Table 9.

TABLE 9 Seashore Trees.

Species	Common Name	Native Area
Albizia lophantha	Cape Wattle	Australia
Araucaria excelsa	Norfolk Island Pine	Norfolk Island
Banksia integrifolia	Coast Honeysuckle	Australia
Barringtonia acutangula		Sri Lanka
Caesalpinia coriaria	Divi-divi	Venezuela
Carallia integerrima	Dawata	India
Casasia clusiaefolia	Seven Year Apple	Florida
Catesbaea parviflora	Lily Thorn	Florida
Cerbera odollam	Gon-kadura	India
Conocarpus erectus	Buttonwood	Florida
Corynocarpus laevigatus	Karaka	New Zealand
Crataegus pubescens	Mexican Hawthorn	Mexico
Cytissus proliferus	Escabon	Canary Islands
Ficus rubiginosa	Rusty Fig	Florida
Garcinia spicata		India
Grevillea banksii		Australia
Griselinia littoralis	Broadleaf	New Zealand
Guettarda speciosa		South Pacific
Holoptelea integrifolia	Indian Elm	India
Juniperus barbadensis	Barbados Red Cedar	West Indies
Leptospermum laevigatum	Coast Tea Tree	Australia
Messerschmidia argentia	Beach Heliotrope	Hawaii
Metrosideros tomentosa	New Zealand Christmas Tree	New Zealand
Myoporum laetum		New Zealand
Olearoa albida	Tree Aster	New Zealand
Pinus halepensis	Aleppo Pine	Mediterranean
Pittosporum crassifolium	Karo	New Zealand
Pomaderris apetala	Tainui	New Zealand
Prunus spinosa	Sloe	Australia
Pseudopanax crassifolium	Lancewood	New Zealand
Torrubia longifolia	Blolly	Florida
Vitex lucens	Puriri	New Zealand
Ximenia americana	Tallowwood	West Indies

SOURCE: Menninger, 1964.

LIQUID FUELS*

A number of countries are pioneering the large-scale use of alcohol fuels. In Brazil, for instance, a country that imported more than 80 percent of its petroleum in 1979, a combination of factors—including the availability of land and labor, a need for liquid fuels,

---

*See also *Alcohol Fuels: Options for Developing Countries*. To order, see p. 135.

TABLE 10 Utilization of Kallar Grass for Biogas Production.

Material	Yield per Hectare per Year
Kallar grass	40 t (green)
Kallar grass	16.8 t (dry)
Methane (0.18 m³/kg dry matter)	3,024 m³
Sludge (0.72 kg/kg dry matter)	12.1 t
Nitrogen in sludge	240 kg
Total Energy	15×10^6 kcal

SOURCE: Malik et al., 1986.

and a strong base in sugarcane production—has led to an ambitious alcohol fuels program. Instead of producing granular sugar for the world market, sugarcane juice is fermented to ethanol. This alcohol is used both in combination with gasoline and as a complete substitute for gasoline in Brazil's automobiles. In Costa Rica, Indonesia, Kenya, Papua New Guinea, the Philippines, Sudan, Thailand, and other countries, alcohol fuel projects are being examined or developed.

The opportunity also exists for the production of liquid fuels from salt-tolerant plants. The sugar beet, *Beta vulgaris*, can be grown with saline water. The techniques for extracting sugar from this crop and fermenting it to ethanol are well known and widely practiced. Although less well known, the nipa palm (*Nypa fruticans*) is also a potential source of sugar for conversion to ethanol.

The nipa palm flourishes in the tidal marshes and on the submerged banks of bays and estuaries from West Bengal through Burma, Malaysia to northern Australia. There are extensive stands in the Philippines, Papua New Guinea, and Indonesia.

Nipa sap contains about 15 percent sugar, which can be collected from the mature fruit stalk after the fruit head has been cut off. Carefully done, tapping can continue for an extended period and considerable quantities of sap can be harvested. Pratt et al. (1913) report yields of 40 liters per tree per season, which they project as 30,000 liters of juice per hectare each year. Cultivated palms may produce as much as 0.46 liters of sap per tree each day, which is equivalent to nearly 8,000 liters of alcohol per hectare each year.

Because of the presence of wild yeasts, the sap begins to ferment as soon as it is tapped; if it is not used quickly, fermentation will proceed to acetic acid. The principal disadvantages for nipa are the inaccessibility of its wild stands and the difficulty of working in the swampy terrain that the plant prefers. Cultivated stands may require land that would otherwise be suitable for rice.

GASEOUS FUELS

Although grown primarily for use as fodder (see p.75), kallar grass (*Leptochloa fusca*) has been shown to have potential as an energy crop by researchers at the Nuclear Institute for Agriculture and Biology in Pakistan. As shown in Table 10, when kallar grass is used as a substrate for biogas production, the energy yield per hectare per year is estimated to be 15×10^6 kcal.

REFERENCES AND SELECTED READINGS

General

Adappa, B. S. 1986. Waste land development for bioenergy need for forestry grant schemes and incentive policies. *MYFOREST* 22(4):227-231.

Ahmad, R. 1987. *Saline Agriculture at Coastal Sandy Belt*. University of Karachi, Karachi, Pakistan.

Barrett-Lennard, E. G., C. V. Malcolm, W. R. Stern and S. M. Wilkins (eds.). 1986. *Forage and Fuel Production from Salt Affected Wasteland*. Elsevier, Oxford, England. (Also published as Volume 5, No. 1-3, 1986, of *Reclamation and Revegetation Research*).

Bangash, S. H. 1977. Salt tolerance of forest tree species as determined by germination of seeds at different salinity levels. Chemistry Branch, Pakistan Forest Institute, Peshawar, Pakistan.

Chaturvedi, A. N. 1984. Firewood crops in areas of brackish water. *Indian Forester* 110(4):364-366.

Goodin, J. R. 1984. Assessment of the Potential of Halophytes as Energy Crops for the Electric Utility Industry (Final Report). International Center for Arid and Semi-Arid Land Studies, Lubbock, Texas, US.

Gupta, G. N., K. G. Prasad, S. Mohan and P. Manivachakam. 1986. Salt tolerance of some tree species at seedling stage. *Indian Forester* 112(2):101-113.

Jambulingam, R. and E. C. M. Fernandes. 1986. Multipurpose trees and shrubs on farmlands in Tamil Nadu State (India). *Agroforestry Systems* 4:17-32.

Le Houérou, H. N. 1986. Salt-tolerant plants of economic value in the Mediterranean basin. *Reclamation and Revegetation Research* 5:319-341.

Lima, P. C. F. 1986. Tree productivity in the semiarid zone of Brazil. *Forest Ecology and Management* 16:5-13.

Malik, M. N. and M. I. Sheikh. 1983. Planting of trees in saline and waterlogged areas. Part I. Test planting at Azakhel. *Pakistan Journal of Forestry* 33(1):1-17.

Menninger, E. A. 1964. *Seaside Plants of the World.* Hearthside Press, Great Neck, New York, US.

Midgley, S. J., J. W. Turnbull and V. J. Hartney. 1986. Fuel-wood species for salt affected sites. *Reclamation and Revegetation Research* 5:285-303.

Morris, J. D. 1983. The role of trees in dryland salinity control. *Proceedings of the Royal Society of Victoria* 95(3):123-131.

Morris, J. D. 1984. Establishment of trees and shrubs on a saline site using drip irrigation. *Australian Forestry* 47(4):210-217.

Negus, T. R. 1984. Trees for saltland. *Farmnote* 67/84. Western Australian Department of Agriculture, South Perth, Australia.

O'Leary, J. W. 1979. The yield potential of halophytes and xerophytes. Pp. 574-581 in: J. R. Goodin and D. K. Northington (eds.) *Arid Land Plant Resources.* Texas Tech University, Lubbock, Texas, US.

Patel, V. J. 1987. Prospects for power generation from waste land in India. *Appropriate Technology* 13(4):18-20.

Sheikh, M. I. 1974. Afforestation in waterlogged and saline areas. *Pakistan Journal of Forestry* 24(2):186-192.

Van Epps, G. A. 1982 Energy biomass from large rangeland shrubs in the Intermountain United States. *Journal of Range Management* 35(1):22-25.

Yadav, J. S. P. 1980. Potentialities of salt-affected soils for growing trees and forage plants. *Indian Journal of Range Management* 1:33-44.

Fuelwood Trees

Prosopis

Almanza, S. G. and E. G. Moya. 1986. The uses of mesquite (*Prosopis* spp.) in the highlands of San Luis Potosi, Mexico. *Forest Ecology and Management* 16:49-56.

Esbenshade, H. W. 1980. Kiawe (*Prosopis pallida*): a tree crop in Hawaii. *International Tree Crops Journal* 1(2/3):125-130.

Felker, P., G. H. Cannell and J. F. Osborn. 1983. Effects of irrigation on biomass production of 32 prosopis (mesquite) accessions. *Experimental Agriculture* 19(2):187-198.

Felker, P., P. R. Clark, A. E. Laag and P. F. Pratt. 1981. Salinity tolerance of the tree legumes mesquite (*Prosopis glandulosa* var. *torreyana*, *P. velutina* and *P. articulata*), algarrobo (*P. chilensis*), kiawe (*P. pallida*) and tamarugo (*P. tamarugo*) grown in sand culture on nitrogen-free media. *Plant and Soil* 61(3):311-317.

Khan, D., R. Ahmad and S. Ismail. 1986. Case history of *Prosopis juliflora* plantation at Makran coast raised through saline water irrigation. Pp. 559-585 in: R. Ahmad and A. San Pietro (eds.) *Prospects for Biosaline Research.* University of Karachi, Karachi, Pakistan,.

Marmillon, E. 1986. Management of algarrobo (*Prosopis alba*, *Prosopis chilensis*, *Prosopis flexuosa*, and *Prosopis nigra*) in the semiarid regions of Argentina. *Forest Ecology and Management* 16:33-40.

Muthana, K. D. and B. L. Jain. 1984. Use of saline water for raising tree seedlings (*Prosopis juliflora*, *Leuceana leucocephala*). *Indian Farming* 34(2):37-38.

Rhodes, D. and P. Felker. 1988. Mass screening of *Prosopis* (Mesquite) seedlings for growth at seawater salinity concentrations. *Forest Ecology and Management* 24(3):169-176.

Eucalyptus

Biddiscombe, E. F., A. L. Rogers, E. A. N. Greenwood and E. S. DeBoer. 1981. Establishment and early growth of species in farm plantations near saline seeps. *Australian Journal of Ecology* 6:383-389.

Biddiscombe, E. F., A. L. Rogers, E. A. N. Greenwood and E. S. DeBoer. 1985. Growth of tree species near salt seeps, as estimated by leaf area, crown volume and height. *Australian Forest Research* 15(2):141-154.

Blake, T. J. 1981. Salt tolerance of eucalypt species grown in saline solution culture. *Australian Forest Research* 11(2):179-183.

Carr, S. G. M. and D. J. Carr. 1980. A new species of *Eucalyptus* from the margins of salt lakes in Western Australia. *Nuytsia* 3:173-178.

Chippendale, G. M. 1973. *Eucalypts of the Western Australian Goldfields*. Australian Government Publishing Service, Canberra, Australia.

Darrow, W. K. 1983. Provenance-type trials of *Eucalyptus camaldulensis* and *E. tereticornis* in South Africa and Southwest Africa: eight-year results. *South African Forestry Journal* 124(3):13-22.

Grunwald, C. and R. Karshon 1983. Variation of *Eucalyptus camaldulensis* from North Australia grown in Israel. Division of Forestry, Agricultural Research Organization, Ilanot, Israel.

Jacobs, M. R. 1981. *Eucalypts for Planting*. FAO Forestry Series No. 11, Rome, Italy.

Karschon, R. and Y. Zohar. 1975. Effects of flooding and of irrigation water salinity on *Eucalyptus camaldulensis* Dehn. from three seed sources. Leaflet No. 54, Division of Forestry, Agricultural Research Organization, Ilanot, Israel.

Mathur, N. K. and A. K. Sharma. 1984. Eucalyptus in reclamation saline and alkaline soils in India. *Indian Forester* 110(1):9-15.

Muthana, K. D., G. V. S. Ramakrishna and G. D. Arora. 1983. Analysis of growth and establishment of *Eucalyptus camaldulensis* in the Indian arid zone. *Annals of Arid Zone Research* 22(1):151-155.

Sands, R. 1981. Salt resistance in *Eucalyptus camaldulensis* Dehn. from three different seed sources. Division of Soils, CSIRO, Glen Osmond, Australia.

Zohar, Y. 1982. Growth of eucalypts on saline soils in the Wadi Arava. *La-Yaaran* 32(1-4):60-64.

Casuarina

Ng, B. H. 1987. The effects of salinity on growth, nodulation and nitrogen fixation of *Casuarina equisetifolia*. *Plant and Soil* 103:123-125.

Turnbull, J. W. 1986. *Casuarina obesa* Pp. 244-245 in: *Multipurpose Australian Trees and Shrubs*. Australian Center for International Agricultural Research, Canberra, Australia.

van der Moezel, P. G., L. E. Watson, G. V. N. Pearce-Pinto and D. T. Bell. 1988. The response of six *Eucalyptus* species and *Casuarina obesa* to the combined effect of salinity and waterlogging. *Australian Journal of Plant Physiology* 15(3):465-474.

Rhizophora

Bunt, J. S., W. T. Williams and H. J. Clay. 1982. River water salinity and the distribution of mangrove species along several rivers in north Queensland. *Australian Journal of Botany* 30(4):401-412.

Chan, H. T. 1987. Mangrove forest management in the ASEAN region. *Tropical Coastal Area Management* 2(3):6-8.

Chan, H. T. and S. M. Nor. 1987. *Traditional Uses of the Mangrove Ecosystem in Malaysia.* UNDP/UNESCO Regional Mangrove Project, New Delhi, India.

de la Cruz, A. A. 1980. Status of mangrove management in Southeast Asia. *BIOTROP* 1980:11-17. Bogor, Indonesia.

Gordon, D. M. 1988. Disturbance to mangroves in tropical-arid Western Australia: hypersalinity and restricted tidal exchange as factors leading to mortality. *Journal of Arid Environments* 15(2):117-146.

Fortes, M. D. 1988. Mangrove and seagrass beds of East Asia: habitats under stress. *AMBIO* 17(3):207-213.

Khan, Z. H. 1977. Management of the principal littoral tree species of the Sundarbans. Forest Research Institute, Chittagong, Bangladesh.

Morton, J. F. 1976. Craft industries from coastal wetland vegetation. Pp. 254-266 in: M. Wiley (ed.) *Estuarine Processes, Vol.1.* Academic Press, New York, New York, US.

Rutzler, K. and C. Feller. 1988. Mangrove swamp communities. *Oceanus* 30(4):16-34.

Snedaker, S. C. and J. G. Snedaker (eds.). 1984. *The Mangrove Ecosystem.* Unipub, New York, New York, US.

Teas, H. J. (ed.). 1984. *Physiology and Management of Mangroves.* Dr. W. Junk Publishers, The Hague, Netherlands.

Tomlinson, P. B. 1986. *The Botany of Mangroves.* Cambridge University Press, New Rochelle, New York, US.

Melaleuca

Cherrier, J. F. 1981. The niaouli (*Melaleuca quinquenervia*) in New Caledonia. *Revue Forestiere Francaise* 33(4):297-311.

van der Moezel, P. G. and D. T. Bell. 1987. Comparative seedling salt tolerance of several *Eucalyptus* and *Melaleuca* species from Western Australia. *Australian Forestry Research* 17:151-158.

Morton, J. F. 1966. The cajeput tree: a boon and an affliction. *Economic Botany* 20:31-39.

Wang, S., J. B. Huffman and R. C. Littel. 1981. Characterization of melaleuca biomass as a fuel for direct combustion. *Wood Science* 13(4):216-219.

Tamarix

Garrett, D. E. 1979. *Investigation of Woody Biomass for Fuel Production in Warm Climate, Non-Agricultural Land Irrigated with Brackish or Saline Water.* Department of Energy, Washington, DC, US.

Singh, B. and S. D. Khanduja. 1984. Wood properties of some firewood shrubs in northern India (*Tamarix dioca, Carissa spinarum, Acacia calycina, Adhatoda vasica, Dedonia viscosa*). *Biomass* 4(3):235-238.

Acacia

Turnbull, J. W. (ed.). 1986. *Australian Acacias in Developing Countries*. ACIAR Proceedings No. 16, Canberra, Australia.

Adhatoda vasica

Chaturvedi, A. N. 1984. Firewood crops in areas of brackish water. *Indian Forester* 110(4):364-366.

Singh, A., M. Madan and P. Vasudevan. 1987. Increasing biomass yields of hardy weeds through coppicing studies on *Ipomoea fistulosa* and *Adhatoda vasica* with reference to wasteland utilization. *Biological Wastes* 19:25-33.

Pongamia pinnata

Krishnamurthi, A. (ed.). 1969. *Pongamia*. Wealth of India VIII:206-211. CSIR, New Delhi, India.

Lakshmikanthan, V. 1978. *Tree Borne Oil Seeds*. Khadi & Village Industry Commission, Pune, India.

Bringi, N. V. and S. K. Mukerjee. 1987. Karanja seed oil. Pp. 143-166 in: N. V. Bringi (ed.) *Non-Traditional Oilseeds and Oils of India*. Oxford and IBH Publishing Co., New Delhi, India.

Butea monosperma

Manjunath, B. L. (ed.). 1948. *Butea*. Wealth of India I:251-252. CSIR, New Delhi, India.

Lakshmikanthan, V. 1978. *Tree Borne Oil Seeds*. Khadi & Village Industry Commission, Pune, India.

Pithecellobium dulce

Krishnamurthi, A. (ed.). 1969. *Pithecellobium*. Wealth of India VIII:140-142. CSIR, New Delhi, India.

Nipa Palm

Davis, T. A. 1986. Nipa palm in Indonesia, a source of unlimited food and energy. *Indonesian Agricultural Research & Development Journal* 8(2):38-44.

Hamilton, L. S. and D. H. Murphy. 1988. Use and management of nipa palm (*Nypa fruticans*, Arecacae): A review. *Economic Botany* 42:206-213.

Paivoke, A. E. A. 1984. Tapping patterns in the nipa palm. *Principes* 28:132-137.

Pratt, D. S., L. W. Thurlow, R. R. Williams and H. D. Gibbs. 1913. The nipa palm as a commercial source of sugar. *The Philippine Journal of Science* 8(6):377-398.

Kallar Grass

Malik, K. A., Z. Aslam and M. Naqvi. 1986. *Kallar Grass—A Plant for Saline Land*. Nuclear Institute for Agriculture and Biology, Faisalabad, Pakistan.

RESEARCH CONTACTS

General

Rafiq Ahmad, Department of Botany, University of Karachi, Karachi 32, Pakistan.
A. N. Chaturvedi, Conservator of Forests, Research and Development Circle, Lucknow, India.
G. N. Gupta, Forest Soil cum Vegetation Survey, Southern Region, Coimbatore, Tamil Nadu, India.
P. C. F. Lima, EMBRAPA-CPATSA, Cx.P. 23, 56300 Petrolina PE, Brazil.
J. D. Morris, Department of Conservation, Forests and Lands, GPO Box 4018, Melbourne, Victoria 3001, Australia.
David N. Sen, Department of Botany, University of Jodhpur, Jodhpur 342001, India.
Lex Thomson, Tree Seed Centre, CSIRO, PO Box 4008, Queen Victoria Terrace, Canberra, ACT 2600, Australia.
M. I. Sheikh, Forestry Research Division, Pakistan Forest Institute, Peshawar, Pakistan.
J. S. P. Yadav, Central Soil Salinity Research Institute, Karnal 132 001, India.

Prosopis

Empresa Pernambucana de Pesquisa Agropecuaria, Av. Gen. San Martin 1371, CP 1022, Bonji, Recife, PE, Brazil.
Peter Felker, Center for Semi-Arid Forest Resources, Texas A&I University, Kingsville, TX 78363, US.
F. Squella, Estacion Experimental La Platina, Instituto de Investigaciones Agropecuarias (INIA), PO Box 5427, Santiago, Chile.
Holger Stienen, Center for International Development and Migration, Bettinastri 62, 6000 Frankfurt, FRG.
D. Khan, Shoaib Ismail, Department of Botany, University of Karachi, Karachi 32, Pakistan.

Eucalyptus

E. A. N. Greenwood, Division of Water Resources, CSIRO, Private Bag, Wembley, W. A. 6014, Australia.
N. K. Mathur, Forest Research Institute and Colleges, Dehra Dun, India.
K. D. Muthana, Central Arid Zone Research Institute, Jodphur 342 003, India.
Paul G. van der Moezel, Department of Botany, University of Western Australia, Nedlands 6009, Australia.
Yehiel Zohar, Department of Forestry, Agricultural Research Organization, Ilanot 42805, Israel.

Casuarina

M. H. El-Lakany, Desert Development Center, American University in Cairo, 113 Sharia Kasr el Aini, Cairo, Egypt.
S. J. Midgley, Division of Forest Research, CSIRO, PO Box 4008, Queen Victoria Terrace, Canberra, ACT, 2600, Australia.

B. H. Ng, Botany Department, University of Queensland, St. Lucia 4067, Australia.
Paul G. van der Moezel, Department of Botany, University of Western Australia, Nedlands 6009, Australia.

Rhizophora

John S. Bunt, Australian Institute of Marine Science, PMB No. 3, Townsville M.C., Queensland 4810, Australia.
Chan Hung Tuck, Forest Research Institute Malaysia, Kepong, Selangor, Malaysia.
A. A. de la Cruz, Department of Biological Sciences, Mississippi State University, Mississippi State, MS 39762, US.
Francis E. Putz, Department of Botany, University of Florida, Gainesville, FL 32611, US.
Klaus Rutzler, Caribbean Coral Reef Ecosystems, Smithsonian Institution, Washington, DC 20560, US.
UNDP/UNESCO Regional Mangroves Project, 15, Jor Bagh, New Delhi 110003, India.

Melaleuca

J. F. Morton, Morton Collectanea, University of Miami, Coral Gables, FL 33124, US.
Paul G. van der Moezel, Department of Botany, University of Western Australia, Nedlands 6009, Australia.

Tamarix

Chihuahua Desert Research Institute, PO Box 1334, Alpine, TX 79830, US.

Acacia

Division of Forest Research, CSIRO, PO Box 4008, Canberra 2600, Australia.
Forestry Division, Agricultural Research Organization, Ilanot, Israel.

Nipa Palm

T. A. Davis, JBS Haldane Research Center, Nagercoil-4, Tamil Nadu, India.
L. S. Hamilton, East-West Center, Honolulu, HI 96848, US.
E. J. Del Rosario, BIOTECH, UPLB, Los Baños, Philippines.

Kallar Grass

K. A. Malik, Nuclear Institute for Agriculture and Biology, PO Box 128, Faisalabad, Pakistan.

3
Fodder

INTRODUCTION

Halophytes have been used as forage in arid and semiarid areas for millennia. The value of certain salt-tolerant shrub and grass species has been recognized by their incorporation in pasture-improvement programs in many salt-affected regions throughout the world. There have been recent advances in selecting species with high biomass and protein levels in combination with their ability to survive a wide range of environmental conditions, including salinity.

Trees and shrubs can be valuable components of grazing lands and can also serve as shelter and complementary nutrient sources to grasses in arid and semiarid areas. Their deep roots serve as soil stabilizers and nutrient pumps and can lower saline water tables. Trees can provide shade for livestock and shrubs can be used as living fences. Leguminous species improve soil quality by fixing nitrogen.

In arid and semiarid zones, trees and shrubs have several advantages over grasses as fodder. They are generally less susceptible to seasonal variation in moisture availability and temperature, and to fire. Usually less palatable than grasses, they can provide reserve or supplementary feed sources.

In Africa, about 60 percent of the meat production and about 70 percent of the milk production is from arid and semiarid environments. It is here that pastures are most severely degraded and where the planting of trees and shrubs may be most helpful. The use of salt-tolerant species in pasture improvement may allow the use of brackish water for irrigation.

In this section, salt-tolerant grasses, shrubs, and trees with potential for fodder use are described.

GRASSES

Kallar Grass

Kallar grass (*Leptochloa fusca*) is a highly salt-tolerant perennial forage that grows well even in waterlogged conditions. Its deep roots help open hardened soils and harbor nitrogen-fixing bacteria. It recovers well from grazing and can also be cut for hay. Pastures can be established from seed, but the use of rooted slips or stem cuttings yields better results.

Kallar grass is widespread in tropical and southern Africa, the Middle East, and Southeast Asia. Although largely indifferent to rainfall levels, it does require almost constant moisture for its roots. It grows best in waterlogged soils, lake or river margins, and on seasonally flooded flats.

In Pakistan, March is the favored time for planting. A reasonable stand of grass develops in a month, with maximum yields during July and August, the monsoon season. Five cuttings can be obtained during the year with a total yield of about 40 tons of green fodder. Even during the winter months (November through February) when the growth of grass is retarded, a single cutting can yield 3 tons per hectare. Even this low yield is valuable in salt-affected areas where winter fodder is scarce. The grass appears palatable to sheep, goats, buffalo, and cattle.

The qualities that allow kallar grass to grow well under adverse conditions also contribute to its ability to compete well in rice fields and in irrigation canals as a weed.

Silt Grass

Silt grass (*Paspalum vaginatum*) occurs naturally on muddy seacoasts, in tidal marshes, and brackish sandy areas of tropical and subtropical regions. Either erect or prostrate, it has tough, creeping roots and forms dense mats. Once well established, it serves as a useful pasture grass, especially in bog and seepage areas that stay wet with salty water. Although quite suitable for grazing, it dries

Kallar grass can be established from seed, but the use of rooted slips or stem cuttings, here being pressed into a flooded field, gives better results. (K.A. Malik)

slowly and turns black when cut for hay. It has been grown with water containing 1.4 percent salts where ample water was applied for leaching to avoid salt accumulation.

This grass has been found in coastal areas of the West Indies, Belize, Costa Rica, Panama, Venezuela, Guyana, Brazil, Ecuador, Chile, and Argentina in the Western Hemisphere as well as in tidal swamps in Senegal, Sierra Leone, and Gabon. It is widely used for revegetation in saline seepage areas in Australia.

The best means of propagation is through roots, runners, or sod; seeding is not effective. The grass is sensitive to herbicides. Since sheep crop the grass closely and prevent runners from colonizing, grazing protection must be provided until bare areas are covered.

Russian-Thistle

Russian-thistle (*Salsola iberica*) is a salt-tolerant annual common in the western United States. It is well adapted to survive under drought conditions, requiring only about half as much water per unit of dry matter produced as alfalfa. The crude protein content

Silt grass is especially useful for revegetating seepage areas that stay wet with salty water. (C.V. Malcolm)

of Russian-thistle is in the 15-20 percent range and the amino acid composition of this protein is quite similar to that of alfalfa. In a study in New Mexico (USA), biomass yields of 10 tons per hectare were demonstrated.

Although salinity tolerance at germination is low, seedlings tolerate brackish water well, and this exposure seems to improve salinity tolerance in the later vegetative and reproductive stages. Moderate salinity levels resulted in improved yields. Table 11 shows some of these data.

Salsola may also find use as an energy crop. The energy content of field-dried *Salsola* is comparable to lignite. It has been successfully compressed into pellets for use as boiler fuel.

Saltgrasses

Distichlis spicata is used as forage for cattle near Mexico City. Grown on 20,000 hectares of salt flats, this may represent the world's largest area devoted to an introduced halophyte. There are distinct seashore and inland ecotypes; the seashore ecotype has been grown with water twice as salty as seawater.

TABLE 11 Yield and Moisture Content of *Salsola iberica* at Five Salinity Levels. (Saline irrigation was initiated six days after planting; harvest was 64 days after planting.)

Irrigation Salinity (dS/m)	Fresh Weight(g)	Dry Weight(g)	Moisture Content(%)
1.3	921.4	179.6	80.7
10.5	1,217.0	279.8	77.1
18.2	972.8	222.4	77.2
26.7	625.6	131.4	79.0
33.2	386.6	75.0	80.7

SOURCE: Fowler et al., 1985.

Another variety of *Distichlis* developed by NyPa, Inc. has growth rates and nutrient characteristics similar to those of alfalfa. Yields of 20 tons per hectare (dry matter) have been reported using irrigation water containing 1-2 percent salts. A perennial that can tolerate both waterlogging and long periods of drought, it appears suitable for use in many hot arid areas where saline water is available for irrigation.

Channel Millet

Channel millet (*Echinochloa turnerana*) is an uncultivated, wild Australian plant. Its most significant feature is that in its native habitat it requires only a single watering to develop from germination to harvest. It is always found in silty clay that cracks deeply when dry and is subjected to sporadic flooding. Sites may remain dry for years, but when flooding occurs growth is abundant. The seed will not germinate after light rains; deep flooding is required.

Channel millet grows almost exclusively in the so-called "channel country" of Queensland in inland Australia, where it is recognized as a productive, palatable, and nutritious fodder grass. The grain is consumed by cattle, horses, and sheep. In addition, the leaves, culms, and seedheads are eaten by livestock and the whole plant makes excellent hay.

Little is known about the agronomy of channel millet; few attempts have been made to domesticate it and there is little documented information on its botany, germination, growth, environmental requirements, and yield. Laboratory salinity testing indicated that a 50-percent reduction in grain yield occurs at 24 dS/m. Some species of *Echinochloa* are ruinous weeds in rice fields. The weediness of channel millet is unknown, but quarantine measures should be used in its testing to prevent inadvertent release.

E. crus-galli is reported to be a good fodder for cattle, with its grain fed in time of scarcity. *E. frumentacea* is grown as a quick-maturing (six weeks) food crop in India. Both are grown in Egypt on lands too saline for other crops.

Cordgrasses

Members of the *Spartina* genus are tough, long-leaved grasses found in tidal marshes in North America, Europe, and Africa. These grasses have hollow stems (culms) and rhizomes. The hollow stems allow air transport from the leaves to the roots during tidal inundation to maintain aerobic conditions in the root zone. Most *Spartina* species propagate vegetatively by means of spreading underground rhizomes, which grow new roots and buds. Seeds are a less important means of propagation for most species. These grasses survive salt water and saline soils by excreting salt through special glands in their leaves.

Spartina alterniflora (smooth cordgrass) is a tall (1-3 m), robust species that grows closest to the water line. It transplants well and can be seeded under some conditions. *S. folioso* (California cordgrass) is shorter (1 m) and produces less seed. It grows along the western coast of North America from California to Mexico. *S. patens* (salt meadow cordgrass) grows densely in marshes in the area of mean high water. *S. patens* has historically been used for grazing or cut for hay.

S. alterniflora tidal salt marshes are important nursery grounds and sources of nutrients for aquatic organisms. These marshes also provide food and habitat for wildlife, reduce shoreline erosion, and assimilate excess nutrients from pollutants such as sewage and agricultural drainage. Because of this, studies have been made of establishment methods and long term stability of man-made marshes. In North Carolina (USA), it was shown that, after four growing seasons, there was no difference in growth between a transplanted *S. alterniflora* marsh and an adjacent natural marsh. Biomass production of the two marshes was similar during the remainder of the ten-year study.

Smooth cordgrass grows well close to the water line in the background. Salt meadow cordgrass grows best in the area of mean high tide. Two rows of salt meadow cordgrass are markedly reduced in growth as they extend seaward into the realm of smooth cordgrass. (E.D. Seneca)

Rhodes Grass

Rhodes grass (*Chloris gayana*) has been grown in the United Arab Emirates to supply fodder for a rapidly growing livestock population. When irrigation water with a salinity level of 6,000 ppm of dissolved salts was used, the survival of seedlings dropped to zero. However, when the grass was started in a nursery and tufts transplanted to the field, normal growth was obtained with water containing a salt load of up to 15,000 ppm. Success was attributed to good soil drainage. No difficulties were encountered in areas with deep sandy soils.

In laboratory work with *C. gayana*, five successive generations were grown on sand irrigated with NaCl solutions up to 0.7 M (about 4.2 percent). The most successful survivors were found to have not only greater salt tolerance but an improved ability to withstand multiple harvests even at salt levels of about 2 percent.

Tall Wheat Grass

Tall wheat grass (*Elytrigia* [*Agropyron*] *elongata*) is native to southern Russia and Asia Minor where it grows in seashore marshes. It was introduced in Australia more than 50 years ago, where it has since been used for revegetating salted areas. A perennial, it is well adapted to poorly drained saline soils. Although it grows moderately well on saline areas that are permanently wet, best growth occurs where the soil dries out in the summer. Tall wheat grass can be established from seed. It germinates well but is slow to establish. Once a crown of stems develops near ground level, it can withstand moderate grazing.

Other Species

Sporobolus airoides (see p. 20), *S. helvolus*, and *S. maderaspatanus* are all grown on sandy and saline soils in India as fodder for horses and cattle. In Pakistan, irrigation of *S. arabicus* with 17 dS/m water gave yields of 3.9 kg per m^2 per year. In recent tests, *S. stapfianus* demonstrated salt tolerance comparable to kallar grass.

Puccinellia distans (North Africa) and *P. ciliata* (Australia) are fodder grasses highly tolerant to salinity. Puccinellia has been widely used on saline areas in Australia. The plant is an outstanding pioneer species on bare salted land. Seedlings grow slowly and establishment is most successful on bare areas where there is no competition from other plants and where there is protection from grazing. Crude protein contents of 4 percent and digestibilities of about 50 percent are common.

Hedysarum carnosum is a biennial fodder legume that occurs in eastern Algeria and Tunisia on saline clay soils. Native stands in southern Tunisia may yield 2,000-3,000 kg dry matter per hectare per year. Data on *H. carnosum* and other salt-tolerant Mediterranean basin forage grasses are shown in Table 12.

SHRUBS

Although shrubs such as the saltbush (*Atriplex*) and bluebush (*Maireana*) occur widely on saline soils, their salt tolerance at germination is poor. *Atriplex* species have relatively narrow temperature ranges under which germination will occur. As the external salt concentration increases, the temperature range for germination narrows. When saltbush and bluebush species are sown on saline soils under

Puccinellia ciliata is highly salt tolerant and is an outstanding pioneer species on bare salted land. Here it is growing in and through salt crystals. (C.V. Malcolm)

natural rainfall conditions, there is a delicate balance between temperature and salinity levels and the germination and establishment of the seedlings. Such salt-tolerant shrubs may be started in nurseries before being planted in potential grazing areas, but this increases the cost of establishment significantly since it is usually less expensive to plant seed than seedlings.

The use of the Mallen Niche Seeder* overcomes some of these problems. In one pass, the seeder performs the following functions:

• Creates two furrows to collect water next to the seed planting site;

• Forms a central ridge to raise the seed above the level of the surrounding area to reduce waterlogging and aid salt leaching;

• Molds a niche on the top of the ridge to give a sheltered depression for the seed and mulch and to collect rain; and

---

*C. V. Malcolm and R. J. Allen. 1981. The Mallen Niche Seeder for plant establishment on difficult sites. *Australian Rangeland Journal* 3:106-109.

TABLE 12 Fodder Grasses Growing on Salt-Affected Land in the Mediterranean Basin.

Species	Rainfall[1]	Frost Tolerance[2]	Salt Tolerance[3] EC dS/m
Non-Legumes			
(Perennial)			
Festuca elatior (subspecies arundinacea)	400	G	20
Elytrigia elongatum	300	G	20
Agropyropsis lolium	300	G	20
Pucciniella distans	200	G	20
Sporobolus tourneuxii	50	F	20
S. helvolus	50	F	20
Legumes			
(Annual & Biannual)			
Medicago ciliaris	400	F	10
M. intertexta	400	F	10
M. hispada	200	F	10
Hedysarum carnosum	150	F	30
Melilotus indica	300	F	10
M. alba	300	G	10
(Perennial)			
Trifolium fragiferum	400	G	15
Lotus creticus	150	P	10
L. corniculatus	400	G	10
Teragonolobus siliquosus	400	G	15

1. Minimum rainfall requirement in mm/yr.
2. Frost tolerance; G = good, F = fair, P = poor.
3. Maximum salt tolerance = electrical conductivity of soil saturation extract at 25°C.
SOURCE: Adapted from Le Houérou, 1986.

- Deposits seed and mulch in the niche at approximately 2 m intervals and sprays the mulch and seed with a black coating to raise the soil temperature.

The seedbed shape, ridge height, and plant spacing can be adjusted for different soil and climatic conditions. In arid areas, the niche is made lower and wider to capture more water; in high rainfall areas, it is made narrower and higher to reduce the danger of waterlogging.

Although newly planted fields can usually be protected from stock animals, the seedlings are attractive to insects, rodents, and other small animals that are more difficult to exclude.

Atriplex

Saltbushes grow throughout the world. They tolerate salinity in soil and water, and many are perennial shrubs that remain green

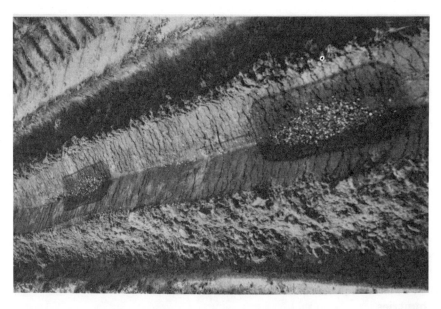

The Mallen Niche Seeder (top) is valuable for establishing shrubs on saline soil. The seeder creates two furrows (bottom) to collect water next to the planting site, forms a niche in the central ridge, deposits seed and mulch in this niche, and sprays the mulch and seed with a black coating to raise the soil temperature. (C.V. Malcolm)

Atriplex undulata plants, sown with the Mallen Niche Seeder in saline soil in Western Australia, are well established at eight months. (C.V. Malcolm)

all year. They are especially useful as forage in arid zones. *Atriplex nummularia*, for example, grows well with only 150-200 mm annual rainfall.

Native stands of *Atriplex* produce about 0.5-4 tons of dry matter per hectare per year. Under rain-fed cultivation, about twice that amount may be obtained. When grown with irrigation, yields equivalent to those of conventional irrigated forage crops can be obtained. And the *Atriplex* can be irrigated with saline water.

Nutritive values for *A. nummularia* and *A. halimus* are high. Both have digestible protein contents averaging near 12 percent of dry matter, about the same as alfalfa. In a year with only 200 mm of rainfall, these two species supported 1,000-1,500 feed units per hectare, about eight times better than a good native pasture under the same conditions. They also survived a year with only 50 mm of rainfall. Although *A. nummularia* has poor palatability, a palatable type has been selected in South Africa. It has been successfully introduced in North and South Africa and several South American countries.

A. canescens (four wing saltbush) is native to semiarid areas of North America where spring and fall rainfall patterns are typical.

An eight-month-old *Atriplex cinerea* grows vigorously on a saline seepage in southwestern Australia. (C.V. Malcolm)

Its nutritive value is as high as *A. nummularia* and it can be seeded in saline soil. Pasture with a mixed population of *A. canescens* and native vegetation sustained three sheep per hectare with 250 mm annual rainfall. *A. canescens* is also palatable to cattle.

In Israel and North Africa, a Mediterranean species, *A. halimus*, has proven hardier than *A. nummularia* or *A. canescens*. Although less palatable, it will grow in shallow soil and on slopes where other plants cannot survive. It does well with a winter rainfall of 200 mm but should be interplanted with more palatable species.

A. patula grows on higher ground and does not tolerate prolonged flooding or immersion in salt water. It has grown well when irrigated with 2.5-3.2 percent saline water, yielding 1.2 tons per hectare of seed with 16 percent crude protein.

A. polycarpa reportedly produces vegetative yields equivalent to alfalfa even when irrigated with water containing 3-4 percent salt. The protein content of *A. polycarpa* is about the same as alfalfa.

A. amnicola (formerly *A. rhagodioides*) is a spreading bush that can reach 4 m in diameter and 1 m in height. Prostrate branches take root to expand coverage. Mulch-covered seeds can be used for

Australian farmers obtain a better return from salinized land by raising sheep on *Atriplex* species than by growing wheat. Here sheep graze *A. undulata* and *A. lentiformis*. (C.V. Malcolm)

introductions in new areas. Once established, it tolerates grazing well. It is particularly suited for waterlogged conditions.

A. amnicola grazed in autumn provided 1,588 sheep-grazing days per hectare (average over 6 years) in a 350-mm rainfall zone of Australia. Heavy grazing failed to damage the stand and many new plants were established. Establishment of *A. amnicola* in saline soils is improved by using genotypes selected for their tendency to produce volunteer plants.

A. undulata, from Argentina, is in widespread use on salt-affected land in Western Australia. Seeds are harvested mechanically and the bushes are established by commercial contractors using direct seeding. *A. undulata* is palatable to sheep, and when used as an autumn reserve feed, provided about 900 sheep-grazing days per hectare in a 300 mm rainfall zone. *A. lentiformis*, from the southwest United States, is included with *A. undulata* sowings on salt-affected soil in southwest Australia.

A. halimus has been grown irrigated with a nutrient solution containing 3.0 percent sodium chloride. Propagation of *A. halimus* is straightforward. Seedlings or cuttings are grown in a nursery for

The only leaves remaining on this *Atriplex nummalaria* are those the sheep cannot reach. (G. Shay)

3-6 months and then planted in the field in early spring, preferably after rain. In Israel, washed seed planted directly into moist soil established well. Grazing should be deferred for two or three years until the plants are about 1.5 m high.

The importance of long-term adaptation studies has been demonstrated in Iran, where extensive plantings of *A. halimus* and *A. lentiformis* suffer from a disease not found in their native habitats. In northeastern Iran, *A. lentiformis* is unable to regenerate from seed, apparently because of the high temperatures required for germination.

About two million *Atriplex* plants are arrayed for transplanting into the rangelands of northeastern Iran. (C.V. Malcolm)

As part of an extensive evaluation of halophytes in Israel, seven *Atriplex* species were grown using 100 percent seawater irrigation. Results of these experiments are shown in Table 13.

Of these seven species, *A. barclayana* is outstanding both in terms of salt tolerance and biomass production. This species has been multiplied from vegetative cuttings to develop plantings for animal feeding trials. *A. lentiformis* also produces large quantities of biomass but has a tendency to become woody. It therefore has the potential for both fodder and fuelwood. *A. lentiformis* and *A. canescens* (subsp. *linearis*) have also given high yields (1.7+ kg per m^2 per year) when grown with hypersaline (about 4 percent total salts) seawater in Mexico's Sonora Desert.

Mairiena

In Australia, there are many *Mairiena* species that are useful for grazing. *Mairiena* are small to medium woody shrubs with succulent leaves and winged, wind-disseminated fruits. In general, they occur in less waterlogged areas than *Atriplex*. *M. brevifolia* is widely grown in Western Australia. It is palatable, recovers well from grazing, and

TABLE 13 Annual Yield and Feed Value of *Atriplex* Species Grown With 100 Percent Seawater Irrigation.

Atriplex Species	Fresh Weight kg/m²	Dry Weight kg/m²	Ash (%)	Fiber (%)	Crude Protein (%)
A. atacamensis	3.75	1.61	23-25	23.2-30.8	9.9-16.5
A. barclayana	8.70	2.09	23-28.5	15.5-22.4	11.9-17.9
A. "camarones"*	4.39	1.51	29.4-37	19.7-29.6	13.8-19.5
A. cinerea	3.90	1.46	28.4-33.5	24.1-30.6	12.6-17.7
A. lentiformis	3.0	2.01	24.0	22.7-27.3	17.6
A. linearis	2.44	1.26	10.5-18.1	24.6-39.5	10.2-14.6
A. undulata	4.50	1.75	24.5-34.2	24.3-30.9	12.6-17.1

*Unidentified *Atriplex* species collected in the region of Camarones, Argentina.

SOURCE: Aronson et al., 1985.

colonizes readily. It has crude protein levels ranging from 15 to 26 percent (dry basis), and serves as a nutritious forage for sheep.

Differences in salt resistance, salt content, drought resistance, leafiness, and palatability have been observed within populations of many of these shrub species. Selection and breeding could greatly improve these characteristics as well as growth habit (to allow easier grazing) and recovery after grazing.

Kochia

Prostrate kochia (*Kochia prostrata*) is a perennial shrub used for browse in Asiatic Russia, where it is consumed by domestic livestock and wildlife. It is well adapted to arid areas and does well on saline and even alkaline soils. Where it has been introduced in the western United States, biomass yields have been good and oxalate levels, a concern with some members of this family, have been low (<2 percent). In a recent test, one accession of *K. prostrata* showed no reduction in dry matter yields at soil salinity levels of 17 dS/m.

Kochia indica and *K. scoparia* have been field tested in Saudi Arabia to determine germination and vegetative yields on salt-affected land using saline water (0.53 percent total dissolved solids) for irrigation. Both *K. indica* and *K. scoparia* germinated well when seeds were planted at <1 cm deep. Irrigated growth from March

Kochia indica has been grown for fodder on salt-affected land in Saudi Arabia. Using water with about 0.5 percent salt for irrigation, yields were 8.5 kg per bush after six months growth. (M.A. Zahran)

through August gave mean fresh weights of 8.5 kg per bush for *K. indica* and 5.6 kg per bush for *K. scoparia*.

Samphire

Samphires are succulent, highly salt-tolerant perennial shrubs that occur naturally on waterlogged saltland throughout agricultural areas in Western Australia. The most common species are black-seeded samphire (*Halosarcia pergranulata*), pale-seeded samphire (*H. lepidosperma*), and woody-seeded samphire (*H. indica* ssp. *bidens*).

Samphire plants do not have true leaves. The stem is thickened into a succulent cylinder with joints at the points where leaves or shoots would normally be. The black-seeded samphire contains about 14 percent crude protein on a dry basis. Pale-seeded samphire is generally lower in protein and higher in salt. Since all samphires contain high levels of salt, excessive salt intake by grazing animals is possible. Water with a low salt content and alternative feeds should be provided.

TREES

Trees can be used as forage in several ways. Trees with low branches can be grazed directly. Management of these stands can involve seasonal control of stocking rates to avoid periods when the plants are susceptible to grazing damage. Trees with branches out of the reach of livestock can provide fallen leaves and pods for fodder. Such taller trees can also be lopped for fodder. Trees of any size can be protected in their stands and fodder cut and carried to the livestock.

Acacia

*Acacia** species are widely used in arid and saline environments as supplementary sources of fodder. Although dry matter digestibility of *Acacia* leaves has not been determined for a large number of species, available data indicate it is relatively low. This is probably associated with the high lignin content of the cell wall and the presence of tannins, both of which inhibit digestibility.

Acacia pods provide food for livestock in large areas of the semiarid zone of Africa. Since most of the *Acacia* branches are above the reach of the livestock, overgrazing is not a problem.

A. cyclops and *A. bivenosa* tolerate salt spray and salinity. They grow on coastal dunes as small trees or bushy shrubs. Pods and leaves of both are consumed by goats. Although salt tolerance is likely in many *Acacia* species found in coastal areas, it is unmeasured or unconfirmed for most.

A. ampliceps grows in saline soils in northwestern Australia and appears to be a useful fodder species. Other Australian *Acacias* with potential for use as fodder include *A. holosericea*, *A. saligna*, *A. salicina*, and *A. victoriae*.**

Leucaena

*Leucaena leucocephala**** is a tree legume widely cultivated in tropical and subtropical countries. It is both salt and drought resistant. Leaves, pods, and seeds are browsed by cattle, sheep, and

---

*See also *Mangium and Other Fast-Growing Acacias for the Humid Tropics*. To order, see p. 135.

**Personal communication, Lex Thomson, CSIRO, Australia

***See also *Leucaena: Promising Forage and Tree Crop for the Tropics*. To order, see p. 135.

The leaves and pods of *Prosopis* species are used as forage for cattle, goats, sheep, and camels throughout the world. Here a camel grazes *P. juliflora* in Rajasthan, India. (J.A. Aronson)

goats. In Pakistan, it has been grown on coastal sandy soil through irrigation with saline (14 dS/m) water. When seawater comprised 20 percent of the irrigation water, yields were reduced by 50 percent.

In the Ryukyu Islands, where monsoon winds carry seawater into windward pastures, salt-tolerant fodder sources are needed. Among nine tropical legumes tested in a forage-production project, *Leucaena* showed the highest salt tolerance.

Prosopis

The leaves and pods of mesquite (*Prosopis* spp.) have been used as forage for cattle, goats, sheep, and camels in countries throughout the world—*P. juliflora* and *P. cineraria* in India, *P. chilensis* in South America, *P. glandulosa* in the United States, and *P. pallida* in Australia.

In the Pampa del Tamarugal of northern Chile, the annual rainfall is less than 50 mm, the water table ranges from 2 to 20 m, and a crust of salt about 0.5 m thick covers much of the ground. About 20 years ago the Chilean government began to improve this area by growing tamarugo (*P. tamarugo*). In some cases, these trees were

Almost 23,000 hectares of tamarugo forest have been established on the salt-encrusted Pampa del Tamarugal of northern Chile. (Instituto Forestal)

planted in pits dug through the salt into the soil. Although watering was required for the first year, after that the plants survived by capturing moisture from the ground and air.

About 23,000 hectares are now covered with tamarugo forest. The trees are 8-15 m in height with trunks up to 35 cm in diameter. After the trees reach about 10 m, further growth is very slow and the tree diverts more and more energy from photosynthesis to the production of leaves and fruits. The leaves are rich in carbohydrates and protein and have a feeding value similar to that of hay. The fruits have 37-61 percent digestibility and are an excellent feed for sheep and goats.

About 1.5 sheep per hectare can subsist on the tamarugo forest range and produce about 3-5 kg of wool per fleece. Supplemental feeding with alfalfa raises meat yields.

In addition, the dense and durable tamarugo wood finds many uses. The heartwood is extremely resistant to weathering and has desirable timber qualities. It is used for heavy construction, railway ties, poles, furniture, tool handles, and, because of its hardness, for parquet floors. It also makes superior firewood and can be used to produce a high quality charcoal as well.

REFERENCES AND SELECTED READINGS

General

Ahmad, R. 1987. *Saline Agriculture at Coastal Sandy Belt*. University of Karachi, Karachi, Pakistan.

Barrett-Lennard, E. G., C. V. Malcolm. W. R. Stern and S. M. Wilkins (eds.). 1986. *Forage and Fuel Production from Salt Affected Wasteland*. Elsevier, Oxford, UK.

Greenwood, E. A. N. 1986. Water use by trees and shrubs for lowering saline groundwater. *Reclamation and Revegetation Research* 5:423-434.

Le Houérou, H. N. 1986. Salt tolerant plants of economic value in the Mediterranean basin. *Reclamation and Revegetation Research* 5:319-341.

Le Houérou, H. N. 1985. Forage and fuel plants in the arid zone of North Africa, the Near and Middle East. Pp. 117-141 in: G. E. Wickens, J. R. Goodin and D. V. Field (eds.) *Plants for Arid Lands*. George Allen & Unwin, London, UK.

Le Houérou, H. N. 1979. Resources and potential of the native flora for fodder and sown pasture production in the arid and semi-arid zones of North Africa. Pp. 384-401 in: J. R. Goodin and D. K. Northington (eds.) *Arid Land Plant Resources*. Texas Tech University, Lubbock, Texas, US.

Looijen, R. C. and J. P. Bakker. 1987. Utilization of different salt-marsh plant communities by cattle and geese. Pp. 52-64 in: A. H. L. Huiskes, C. W. P. M. Blom and J. Rozema (eds.) *Vegetation Between Land and Sea*. W. Junk Publishers, Dordrecht, Netherlands.

Mahmood, K., K. A. Malik, K. H. Sheikh and M. A. K. Lodhi. 1989. Allelopathy in saline agricultural land: vegetation successional changes and patch dynamics. *Journal of Chemical Ecology* 15(2):565-579.

McKell, C. M. 1986. Propagation and establishment of plants on arid saline land. *Reclamation and Revegetation Research* 5:363-375.

Rautenstrauch, K. R., P. R. Krausman, F. M. Whiting and W. H. Brown. 1988. Nutritional quality of Desert Mule Deer forage in King Valley, Arizona. *Desert Plants* 8(4):172-174.

Sen, D. N., R. B. Jhamb and D. C. Bhandari. 1985. Utilization of saline areas of Western Rajasthan through suitable plant introduction. *GEOBIOS* 1985:348-360.

Zedler, J. B. 1984. *Salt Marsh Restoration*. California Sea Grant Program, University of California, La Jolla, California 92093, US.

Grasses

Kallar Grass

Malik, K. A., Z. Aslam and M. Naqvi. 1986. *Kallar Grass—A Plant for Saline Land.* Nuclear Institute for Agriculture and Biology, Faisalabad, Pakistan.

Qureshi, R. H., M. Salim, M. Abdullah and M. G. Pitman. 1982. *Diplachne fusca*: an Australian salt-tolerant grass used in Pakistani agriculture. *Journal of the Australian Institute of Agricultural Science.* 48:195-199.

Sandhu, G. R., Z. Aslam, M. Salim, A. Sattar, R. H. Qureshi, N. Ahmad and R. G. Wyn Jones. 1981. The effect of salinity on the yield and composition of *Diplachne fusca* (kallar grass). *Plant, Cell and Environment* 4:177-181.

Silt Grass

Anonymous. 1980. Salt-water couch—for salty seepages and lawns. *Farmnote* 23/80. Western Australia Department of Agriculture, South Perth, Australia.

Morton, J. F. 1973. Salt-tolerant silt grass (*Paspalum vaginatum*). *Proceedings of the Florida State Horticultural Society* 86:482-490.

Russian-Thistle

Foster, K. E., M. M. Karpisak, J. G. Taylor, and N. G. Wright. 1983. Guayule, Jojoba, Buffalo Gourd and Russian Thistle: Plant Characteristics, Products and Commercialization Potential. *Desert Plants* 5(3):112-126.

Fowler, J. L., J. H. Hageman, and M. Suzukida. 1985. *Evaluation of the Salinity Tolerance of Russian Thistle to Determine its Potential for Forage Production using Saline Irrigation Water.* New Mexico Water Resources Institute, Las Cruces, New Mexico, US.

Saltgrasses

Wrona, A. F. and E. Epstein. 1982. Screening for salt tolerance in plants: an ecological approach. Pp. 559-564 in: A. San Pietro (ed.) *Biosaline Research.* Plenum Press, New York, New York, US.

Channel Millet

Shannon, M. C., E. L. Wheeler and R. M. Saunders. 1981. Salt tolerance of Australian channel millet. *Agronomy Journal* 73:830-832.

Sastri, B. N. (ed.). 1952. *Echinochloa. The Wealth of India* III:124-126. CSIR, New Delhi, India.

Cordgrasses

Broome, S. W., E. D. Seneca and W. W. Woodhouse, Jr. 1986. Long-term growth and development of transplants of the salt-marsh grass *Spartina alterniflora. Estuaries* 9(1):63-74.

Rhodes Grass

Malkin, E. and Y. Waisel. 1986. Mass selection for salt resistance in Rhodes grass (*Chloris gayana*). *Physiologica Plantarum* 66:443-446.

Tariq, A. R. and H. M. A. Tayab. 1984. Cultivation of *Chloris gayana* cv: Pioneer on saline water under hyper-arid climate. *The Pakistan Journal of Forestry* 34(7):151-154.

Tall Wheatgrass

Roundy, B. A. 1985. Emergence and establishment of basin wildrye and tall wheatgrass in relation to moisture and salinity. *Journal of Range Management* 38(2):126-131.

Shannon, M. C. 1978. Testing salt tolerance among tall wheatgrass lines. *Agronomy Journal* 70:719-722.

Hedysarum carnosum

Le Houérou, H. N. 1986. Salt tolerant plants of economic value in the Mediterranean basin. *Reclamation and Revegetation Research* 5:319-341.

Puccinellia

Negus, T. R. 1982. Puccinellia—its grazing value and management. *Farmnote* 34/82. Western Australia Department of Agriculture, South Perth, Australia.

Negus, T. R. 1980. Spray-seed for puccinellia establishment. *Farmnote* 17/80. Western Australia Department of Agriculture, South Perth, Australia.

Sporobolus

Ahmad, R. 1987. *Saline Agriculture at Coastal Sandy Belt*. University of Karachi, Karachi, Pakistan.

Chadha, Y. R. (ed.). 1976. *Sporobolus*. *The Wealth of India* X:24-25. CSIR, New Delhi, India.

Wood, J. N. and D. F. Gaff. 1989. Salinity studies with drought-resistant species of *Sporobolus*. *Oecologia* 78:559-564.

Shrubs

General

Malcolm, C. V. and T. C. Swaan. 1989. *Screening Shrubs for Establishment and Survival on Salt-affected Soils in Southwestern Australia*. Technical Bulletin 81. Western Australia Department of Agriculture, South Perth, Australia.

Malcolm, C. V. 1986. Saltland management-revegetation. *Farmnote* 44/86. Western Australia Department of Agriculture, South Perth, Australia.

Malcolm, C. V. 1983. Seeding shrub pastures on saltland. *Farmnote* 43/83. Western Australia Department of Agriculture, South Perth, Australia.

Otsyina, R., C. M. McKell and G. Van Epps. 1982. Use of range shrubs to meet nutrient requirements of sheep grazing on crested wheatgrass during fall and early winter. *Journal of Range Management* 35(6):751-753.

Atriplex

Aronson, J. A., D. Pasternak and A. Danon. 1985. Introduction and first evaluation of 120 halophytes under seawater irrigation. in: E. E. Whitehead, C. F. Hutchinson, B. N. Timmermann and R. G. Varady (eds.) *Arid Lands Today and Tomorrow*. Westview Press, Boulder, Colorado, US.

El Hamrouni, A. 1986. *Atriplex* species and other shrubs in range improvement in North Africa. *Reclamation and Revegetation Research* 5:151-158.

Franclet, A. and H. N. Le Houérou. 1971. *Les Atriplex en Tunisie et en Afrique du Nord*. FAO, Rome, Italy.

Glenn, E. P. and J. W. O'Leary. 1985. Productivity and irrigation requirements of halophytes grown with seawater in the Sonoran Desert. *Journal of Arid Environments* 9:81-91.

Malcolm, C. V. 1985. Production from salt affected soils. *Reclamation and Revegetation Research* 5:343-361.

O'Leary, J. W. 1986. A critical analysis of the use of *Atriplex* species as crop plants for irrigation with highly saline water. Pp. 415-432 in: R. Ahmad and A. San Pietro (eds.) *Prospects for Biosaline Research*. University of Karachi, Karachi, Pakistan.

Mairiena

Kok, B., P. R. George and J. Stretch. 1987. Saltland revegetation with salt-tolerant shrubs. *Reclamation and Revegetation Research* 6:25-31.

Malcolm, C. V. 1983. Collecting and treating bluebush seed. *Farmnote* 44/83. Western Australia Department of Agriculture, South Perth, Australia.

Kochia

Davis, A. M. 1979. Forage quality of prostrate kochia compared with three browse shrubs. *Agronomy Journal* 71:822-825.

François, L. E. 1986. Salinity effects on four arid zone plants (*Parthenium argentatum, Simmondsia chinensis, Kochia prostrata* and *Kochia brevifolia*). *Journal of Arid Environments* 11:103-109.

Zahran, M. A. 1986. Forage potentialities of *Kochia indica* and *K. scoparia* in arid lands with particular reference to Saudi Arabia. *Arab Gulf Journal of Scientific Research* 4(1):53-68.

Samphire

Malcolm, C. V. and G. J. Cooper. 1982. Samphire for waterlogged saltland. *Farmnote* 4/82. Western Australia Department of Agriculture, South Perth, Australia.

Trees

Acacia

Goodchild, A. V. and N. P. McMeniman. 1987. Nutritive value of *Acacia* foliage and pods for animal feeding. Pp. 101-106 in: J. W. Turnbull (ed.) *Australian Acacias in Developing Countries*. ACIAR Proceedings No.16, Canberra, Australia.

Turnbull, J. W. 1986. *Acacia ampliceps*. *Multipurpose Australian Trees and Shrubs* (Pp. 96-97). Australian Centre for International Agricultural Research, Canberra, Australia.

Leucaena

Ahmad, R. 1987. *Saline Agriculture at Coastal Sandy Belt*. University of Karachi, Karachi, Pakistan.

Kitamura, Y. 1988. Leucaena for forage production in the Ryukyu Islands. *Japan Agricultural Research Quarterly* 22(1):40-48.

Prosopis

Almanza, S. G. and E. G. Moya. 1986. The use of mesquite (*Prosopis* spp.) in the highlands of San Luis Potosi, Mexico. *Forest Ecology and Management* 16:49-56.

Corporacion de Fomento de la Produccion. 1983. Actividades Forestales y Ganaderas en la Pampa del Tamarugal (1963-1982). Tomo I: Aspectos Forestales y Ganaderas. Tomo II: Aspectos Ganaderos. Tomo III: Aspectos Economicos y Evaluacion Social.

Harden, M. L. and R. Zolfaghari. 1988. Nutritive composition of green and ripe pods of honey mesquite (*Prosopis glandulosa*, Fabaceae). *Economic Botany* 42:522-532.

Lyon, C. K., M. R. Gumbmann and R. Becker. 1988. Value of mesquite leaves as forage. *Journal of the Science of Food and Agriculture* 44(2):111-117.

Marangoni, A. and I. Alli. 1988. Composition and properties of seeds and pods of the tree legume *Prosopis juliflora*. *Journal of the Science of Food and Agriculture* 44(2):99-110.

Stienen, H. 1985. *Prosopis tamarugo* in the Chilean Atacama - ecophysiological and reforestation aspects. Pp. 103-116 in: G. E. Wickens, J. R. Goodin and D. V. Field (eds.) *Plants for Arid Lands*. George Allen & Unwin, London, UK.

Vercoe, T. K. 1987. Fodder potential of selected Australian tree species. Pp. 95-100: in J. W. Turnbull (ed.) *Australian Acacias in Developing Countries*. ACIAR Proceedings No. 16, Canberra, Australia.

Zelada, L. and P. Joustra. 1983. Ganderia en La Pampa del Tamarugal. Panel VI Seminario Desarrollo de Zonas Deserticos de Chile. CORFO, Gerencia de Desarrollo AA 83/45. Santiago, Chile.

RESEARCH CONTACTS

General

James F. O'Leary, University of Arizona, Tucson, AZ 85719, US.
J. P. Bakker, Department of Plant Ecology, University of Groningen, PO Box 14, 9750 AA Haren(Gn), Netherlands.
H. N. Le Houérou, CEPE/Louis Emberger, BP 5051, Montpellier-Cedex 34033, France.
C. M. McKell, School of Natural Sciences, Weber State College, Ogden, UT 84408, US.

Grasses

Kallar Grass

K. A. Malik, Nuclear Institute for Agriculture and Biology, PO Box 128, Faisalabad, Pakistan.
B. Myers, Institute for Irrigation and Salinity Research, Ferguson Road Private Bag, Tatura, Victoria 3616, Australia.
R. G. Wyn Jones, Center for Arid Zone Studies, University College of North Wales, Bangor, Wales, LL57 2UW, UK.

Silt Grass

C. V. Malcolm, Western Australia Department of Agriculture, South Perth 6151, Australia.
J. F. Morton, Morton Collectanea, University of Miami, Coral Gables, FL 33124, US.

Russian-Thistle

J. L. Fowler, Department of Crop and Soil Sciences, New Mexico State University, Las Cruces, NM 88003, US.

Saltgrasses

N. Yensen, NyPa, Inc., 727 North Ninth Avenue, Tucson, AZ 85705, US.

Channel Millet

M. C. Shannon, USDA Salinity Research Laboratory, 4500 Glenwood Drive, Riverside, CA 92501, US.

Cord Grasses

D. L. Drawe, Welder Wildlife Foundation, Sinton, TX 78387, US.
J. L. Gallagher, College of Marine Studies, University of Delaware, Lewes, DE 19958, US.
E. D. Seneca, North Carolina State University, Raleigh, NC 27695, US.

Rhodes Grass

Y. Waisel, Department of Botany, Tel Aviv University, Tel Aviv 69978, Israel.

Tall Wheatgrass

J. Dvorak, Department of Agronomy and Range Science, University of California, Davis, CA 95616, US.
B. A. Roundy, School of Renewable Natural Resources, University of Arizona, Tucson, AZ 85721, US.

Hedysarum carnosum

H. N. Le Houérou, CEPE/Louis Emberger, BP 5051, Montpellier-Cedex 34033, France.

Puccinellia

C. V. Malcolm, Western Australia Department of Agriculture, South Perth 6151, Australia.

Shrubs

Atriplex

R. K. Abdul-Halim, Department of Land Reclamation, Center for Agriculture and Water Resources, Council for Scientific Research, PO Box 2416, Baghdad, Iraq.
A. El Hamrouni, Institut des Regions Arides, 4119 Medenine, Tunisia.
H. N. Le Houérou, CEPE/Louis Emberger, BP 5051, Montpellier-Cedex 34033, France.
C. V. Malcolm, Western Australia Department of Agriculture, South Perth 6151, Australia.
C. M. McKell, School of Natural Sciences, Weber State College, Ogden, UT 84408, US.
J. F. O'Leary, University of Arizona, Tucson, AZ 85719, US.
M. K. Sankary, Range and Arid Zone Ecology Research Unit, University of Aleppo, PO Box 6656, Aleppo, Syria.
D. N. Ueckert, Texas A&M University Research and Extension Center, San Angelo, TX 76901, US.

Maireana

P. R. George, Western Australia Department of Agriculture, South Perth 6151, Australia.
B. Kok, Department of Agriculture, Carnarvon, W. A. 6701, Australia.

Kochia

M. A. Zahran, Department of Botany, Mansoura University, Mansoura, Egypt.

Samphire

C. V. Malcolm, Western Australia Department of Agriculture, South Perth 6151, Australia.

Trees

Acacia

A. V. Goodchild, Division of Animal Sciences, University of Queensland, St. Lucia, 4067, Australia.
Lex Thomson, Tree Seed Centre, CSIRO, PO Box 4008, Queen Victoria Terrace, Canberra, ACT 2600, Australia.
J. W. Turnbull, Australian Centre for International Agricultural Research, GPO Box 1571, Canberra, ACT 2601, Australia.
T. K. Vercoe, CSIRO, PO Box 4008, Queen Victoria Terrace, Canberra, ACT 2600, Australia.

Leucaena

R. Ahmad, Department of Botany, University of Karachi, Karachi 32, Pakistan.
Indian Grassland and Fodder Research Institute, Jhansi, Uttar Pradesh 286003, India.
Centro International de Agricultura Tropical, A.A. 6713, Cali, Colombia.
NifTAL Project, PO Box 0, Paia, Maui, HI 96779, US.

Prosopis tamarugo

Estacion Experimental La Platina, Instituto de Investigaciones Agropecuarias (INIA), PO Box 5427, Santiago, Chile.
Instituto Forestal, Huerfanos 554, Casilla 3085, Santiago, Chile.
P. Joustra, SACOR Ltda., Matias Cousino No. 64, Piso 3, Santiago, Chile.
Holger Stienen, Center for International Development and Migration, Bettinastri 62, 6000 Frankfurt, FRG.

4
Fiber and Other Products

INTRODUCTION

Salt-tolerant plants can be used to produce economically important materials such as essential oils, gums, oils, and resins, pulp and fiber, and bioactive compounds. Further, salt-tolerant plants can be cultivated for landscape use and irrigated with saline water, thereby conserving fresh water for other uses.

ESSENTIAL OILS

Kewda

The male flowers of kewda (*Pandanus fascicularis*), a common species of screw pine in India, are used to produce perfume and flavoring ingredients. The flowers are charged to a copper still, water added, and the mix distilled. The steam and essential oil (approximately 0.3 percent by weight of the flowers) are condensed to produce kewda water, or the kewda vapors are captured in sandalwood oil and formulated from this base. The kewda plant is salt tolerant and has been planted in coastal areas to check drifting sand. Propagation is through suckers or stem cuttings, and flowering starts 3-4 years after planting. An annual income of US$8 per plant has been estimated.

The salt-tolerant kewda (*Pandanus fascicularis*) grows from 2 to 3 m high in coastal India. Kewda flowers are used to produce perfume and flavoring ingredients. (P.K. Dutta)

Mentha and Other Species

In India, economic yields of a number of essential oils were obtained from plants grown on saline alkaline soil. Two *Mentha* species were evaluated. *M. piperita* (for peppermint oil) and *M. arvensis* (for menthol) both gave yields comparable to those obtained on normal soil. Other plants giving satisfactory yields on saline alkaline soil included *Matricaria chamomilla*, *Vetiveria zizanioides*, *Cymbopogon nardus* and *C. winterianus* (for citronella oil), *Tagetes minuta*, *Ocimum kilimandscharicum*, and *Anethum graveolens* (English dill). Palamarosa grass (*C. martinii*), a commercially important essential oil plant, is also reported to grow under moderately saline conditions.

The production of essential oils to provide a new source of income in rural areas is one of the objectives of the Ciskei Essential Oils Project, established in southern Africa in 1972. Other objectives include the modification of extraction techniques for use in rural areas, and the identification of markets for the oils produced.

This effort was planned to provide an income for rural dwellers that was derived from familiar, indigenous resources and required relatively little capital. Over the past four years, this project has produced and exported US$1 million worth of essential oils and provided employment to hundreds of rural dwellers during the harvest season. Although the use of salt-tolerant plants is not the focus of the Ciskei Project, the principles could be applied to provide employment and income in areas where saline water or soil occurs.

GUMS, OILS, AND RESINS

Sesbania bispinosa, commonly known as *dhaincha* in India, is an important legume and fodder crop. It is an erect, multibranched annual, about 2.5 m tall at maturity that grows readily on alkaline saline soils. Often grown for use as a green manure (about 12 tons per hectare), its stalks are sources of fiber and fuel, and the seeds yield a galactomannan gum that can be used for sizing and stabilizing purposes. The seed meal can be used for poultry and cattle feed. *S. sesban* and *S. speciosa* are salt-tolerant perennials used as green manure. *S. sesban* can tolerate waterlogging and salt concentrations of 1.0 percent as a seedling and 1.4 percent as it matures.

Grindelia camporum is a 0.5-1.5 m resinous perennial shrub. It exudes large amounts of aromatic resins that cover the surface of the plant. The resins are nonvolatile mixtures of bicyclic terpene acids, esters, and related structures that are insoluble in water but soluble in organic solvents. The amount of resin produced ranges from 5 to 18 percent of the dried biomass.

The plant appears to be salt tolerant; populations are found in saline flats and near salt lakes and springs. Several species of *Grindelia* occur along the North American Pacific Coast in estuaries or salt marsh habitats. These include *G. humilis*, *G. stricta*, *G. latifolia*, and *G. integrifolia*. All produce diterpene acid resins.

Grindelia resins have properties similar to the terpenoids in wood and gum rosins, which are used commercially in adhesives, varnishes, paper sizings, printing inks, soaps, and numerous other industrial applications (Figure 3). With increasing costs and declining supplies of these wood-based materials, substitutions with *Grindelia* resins in this market (700,000 tons per year) may become practical.

The creosote bush (*Larrea tridentata*) grows over large areas of the Chihuahua, Sonora, and Mojave deserts of North America. Evaluations of *Larrea* resins have shown potential uses as an antioxidant for rubber, as an antifungal agent for agricultural applications, and as a reactive material for polymerization with formaldehyde.

Sapium sebiferum, the Chinese tallow tree, is a small marshland

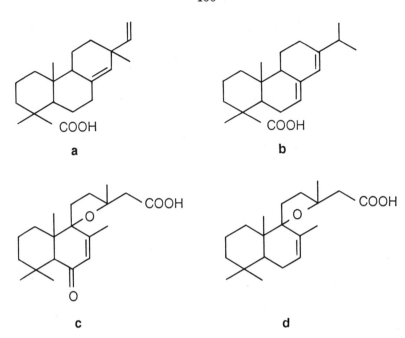

FIGURE 3 Diterpene Acids. *Grindelia* resins have properties similar to wood rosins, which are used in a wide variety of industrial applications. Diterpene acids from *Pinus* (a,b) and *Grindelia* (c,d) are remarkably similar. SOURCE: B.N. Timmerman and J.J. Hoffmann, 1986.

tree native to subtropical China. It has been cultivated there for more than 1,000 years as a source of specialty oils, medicines, and vegetable dyes. The Chinese tallow tree possesses several valuable characteristics: it can be seeded directly; it grows rapidly in warm, waterlogged saline soils; and it resprouts readily.

The major economic potential for this tree is in its high yield of oilseed—more than 10 tons per hectare according to the USDA. The seed contains both an edible hard vegetable fat and an inedible liquid oil, which comprise 45-50 percent of its weight. These oils are physically separated in the seed and may be isolated separately. The edible fat is a potential substitute for cocoa butter and the inedible oil (stillingia oil) appears promising as a drying oil for paints and varnishes. Of the total lipid content in the seed, 30-50 percent is the edible fat.

The seed meal, after extraction of the oil, has a high protein content. It can be used for feed or, with suitable treatment, for human

Rubber rabbitbush grows on saline soils in western North America. It contains natural rubber, a hydrocarbon resin, and constituents that are potential pesticides. (D.J. Weber)

consumption. Five years after planting, when seed production begins, a net return of US$3,200 per hectare per year has been estimated.

Jojoba* (*Simmondsia chinensis*) is a perennial desert shrub with seeds that contain a unique oil. About half of the seed's weight is an oil with a structure similar to sperm whale oil—an ester of a C_{20-22} straight chain alcohol with a C_{20-22} straight chain acid. Both the alcohol and the acid have a terminal double bond, providing a readily accessible site for diverse chemical reactions. This oil and its derivatives have been used primarily in cosmetics, but broader use as a component in specialty lubricants and waxes will probably develop when increased oil production brings lower prices. Currently there are about 16,000 hectares of jojoba plantations in the southwestern United States and other plantations in Mexico, Australia, Israel, Argentina, and South Africa and other African nations.

---

*See also *Jojoba: New Crop for Arid Lands, New Crop for Industry*. To order, see p. 135.

Jojoba is relatively salt tolerant. In California, plants are growing satisfactorily with water containing 0.2 percent salts. In laboratory testing, one variety of jojoba showed no reduction in flower production with 0.6 percent salt. In Israel, jojoba is growing well near the Dead Sea irrigated with brackish water (5-6 dS/m).

While natural rubber occurs in over 2,000 plant species, the commercial source is the rubber tree, *Hevea brasiliensis*. Natural rubber consists of cis-1,4-polyisoprene units. It is preferred in applications that require elasticity, resilience, tackiness, and low heat buildup. It is indispensable for bus, truck, and airplane tires. In 1980, the United States imported about 700,000 tons of natural rubber; imports of about 1 million tons are anticipated for 1990.

Rubber rabbitbush (*Chrysothamnus nauseosus*) is a common desert shrub native to western North America. It grows under a wide range of environmental conditions from Mexico to Canada, commonly appearing on disturbed sites and saline soil. In addition to its forage value, it contains natural rubber and a hydrocarbon resin, and it has constituents that are potential insecticides and fungicides.

The perennial desert shrub, guayule (*Parthenium argentatum*), has also been used as a source of natural rubber.* In 1944, there were 12,000 hectares of guayule planted in California (USA) for rubber production. In tests with guayule, total rubber yields first increased and then decreased as soil salinity increased (Table 14). Interest has recently been revived in guayule for natural rubber.

More recent reports on guayule (Hoffman et al., 1988; Maas et al., 1988) indicate the root-zone salt-tolerance threshold to be about 7.5 dS/m; above this, rubber production is reduced 6.1 percent per unit increase of soil salinity.

Rubber samples from *Hevea*, *Parthenium*, and *Chrysothamnus* appear to be structurally identical. Rubber contents as high as 6.5 percent for *Chrysothamnus* have been reported. If rubber yields of 2 percent are assumed, a plantation would produce 370 kg per hectare after 6 years' growth (guayule yields from California production were higher—about 1,000 kg per hectare after 2 years; *Hevea* yields are about 1,300 kg per hectare per year). Resin contents as high as 21 percent have been reported for *Chrysothamnus*, and some of its hydrocarbon components may find use as insecticides and fungicides.

---

*See also *Guayule: An Alternative Source of Natural Rubber*. To order, see p. 135.

TABLE 14 Plant Growth and Rubber Content of One-Year-Old Guayule Plants at Three Soil Salinities.

Soil Salinity dS/m	Plant Height cm	Fresh Top Weight g	Rubber %
3.2	53	397	3.32
8.7	52	388	6.05
13.2	44	286	4.61

SOURCE: François, 1986.

Compared with guayule, *Chrysothamnus* has several advantages as a potential source of natural rubber. Guayule generally requires good soil, good moisture conditions, and good horticultural practices. Guayule must be grown in frost-free areas because freezing kills it. In contrast, *Chrysothamnus* grows on poor soil, on disturbed sites, and on saline soil. It is found from the hot desert of Arizona to the western arid regions of Canada; there are subspecies that grow at sea level and others that grow at 3,000 m. The rubber content of these plants is similar to that of guayule in natural populations.

PULP AND FIBER

Phragmites australis, common reed, is an ancient marsh plant that has served in roofing, thatching, basketry, and fencing, as well as for fuel. It grows throughout the world in areas with saturated soils or standing water 2.5 m deep or less. The water can be fresh or moderately saline. Nearly any soil from peat to sand is tolerated. Little data exist for yields from managed stands. In the harvest of natural stands, however, productivity is consistently estimated to be about 10 dry tons per hectare. There is current use and broader interest in the manufacture of paper and other cellulose derivatives from this plant.

In Romania, 125,000 tons of *Phragmites* are harvested in the Danube delta each year for use in papermaking. The pulp from these reeds is blended with wood pulp to give a stronger final product.

In Sweden, extensive stands of *Phragmites* have been suggested as an alternative fuel for winter heating. This reed has about 40 percent (by weight) of the energy content of heating oil.

In Egypt, two rushes, *Juncus rigidus* and *J. acutus*, have been

TABLE 15 Germination of *Juncus* spp. With Increasing Salinity.

Germination % at 25°C	NaCl %			
	0.1	1.0	2.0	3.0
J. rigidus	100	100	95	63
J. acutus	15	5	0	0

SOURCE: Zahran and El Demerdash, 1984.

investigated with particular emphasis on their potential use in papermaking. In pilot-level testing, the strength properties of unbleached *J. rigidus* pulp were found to be 73 percent of the kraft pulp ordinarily used. In similar tests, rice straw and bagasse pulps gave only 24 and 42 percent of the strength of kraft pulp.

In germination and propagation testing, *J. rigidus* was much more salt-tolerant than *J. acutus*. Germination results are shown in Table 15.

When rhizomes of these two species were planted on saline test plots, the vegetative yield of *J. rigidus* was almost twice that of *J. acutus*. In studies of the effects of nitrogen and phosphorus fertilizers, vegetative yields and fiber lengths in both species were improved. Increased fiber lengths are an indicator of improved performance in papermaking.

J. rigidus has also been introduced to India from Egypt. Germination, seedling growth, and evaluation of nine-month culms in India indicate that 1.5-2.0 tons per hectare of pulp for papermaking can be produced on saline soil.

The textile screw pine, *Pandanus tectorius*, abounds in tidal flats of Southeast Asia, Malaysia, and Polynesia. The leaves are traditionally and widely used for thatching and basketry. They are also used to fashion wallpaper and lampshades.

Esparto grass (*Stipa tenacissima*) grows in semiarid areas of North Africa. It covers more than 7 million hectares in Algeria and 1.5 million hectares in Tunisia. It has been used for more than a century in papermaking. The paper produced from this fiber is smooth, opaque, and resilient. A paper mill in central Tunisia produces more than 70,000 tons of pulp and paper from this grass, and 20,000 rural families find seasonal work harvesting the crop. In addition, a vegetable wax extracted from the grass before pulping can be used as a substitute for carnauba wax.

In Israel, good yields of cotton have been obtained on sandy soil using drip irrigation with saline water (7.3 dS/m). (G. Shay)

The southern cattail, *Typha domingensis*, is the only *Typha* species that does well in brackish water. It flourishes along the eastern coast of the United States, the Caribbean Islands, and tropical America and Oceania. Its leathery leaves are woven into durable baskets and mats, and are also used for chair seats and backs.

In Pakistan, irrigation with 17 dS/m water gave yields of 5.5 kg per m^2 per year of *Saccharum griffithii*. The roots of this grass are used for rope and mats, and the leaves are used to produce paper pulp.

Kenaf, *Hibiscus cannabinus*, is an annual native to east-central Africa with a fiber content of about 35 percent. Yields of 6-10 tons of dry fiber per acre are possible in five months. Although the projected use of the fiber in the southern United States is for producing newsprint, a major current product from the plant is cordage, which is used for carpet pads, twine, rope, and fiber bags.

In recent work at the U.S. Salinity Laboratory, kenaf was irrigated with water having ECs of up to 6.0 dS/m. Vegetative growth was unaffected by irrigation water salinity up to 4.6 dS/m. Each unit increase above 4.6 dS/m reduced vegetative yield by 36 percent.

Hibiscus tiliaceus, a shrub or small tree, is found near sea shores,

The American oil palm, *Elaeis oleifera*, is found in coastal swamp forests from the lower basin of the Amazon to southern Mexico. Its fruits are a source of oil and tallow, similar to that obtained from the African oil palm. The tree has a low growing habit, which eases fruit harvest. (M.J. Balick)

mangrove swamps, and tidal streams throughout the tropics. The bark fiber is used for ropes, fishing lines, and nets.

Urochondra setulosa is a halophytic grass of the Indus delta and saline marsh flats of the Pakistan coast. This plant dominates sites with ECs of 34-62 dS/m. It merits evaluation as a fiber source.

Cotton (*Gossypium hirusutum*) production using saline water has been examined in the United States, India, Israel, and Tunisia. In Israel, using drip irrigation and four levels of water quality (EC = 1.0, 3.2, 5.4, and 7.3 dS/m), salinity did not reduce yields even at the highest level. In the United States, cotton was drip irrigated with 8.5 dS/m saline water on saline soil in the presence of a saline water table. The yields were equal to that of a control plot that was irrigated with fresh water.

In India, three cotton varieties were reduced in growth and yield when irrigated with seawater diluted to 10,000 and 15,000 ppm salts. In Tunisia, two varieties of cotton were grown with irrigation water containing 0.25, 1.43, 2.43, and 3.45 g per liter of soluble salts. Yield increases of 30-34 percent were obtained at the highest salt level.

The neem tree (*Azadirachta indica*) can be established on poor soils unsuitable for food crops. Neem seed extracts are effective in the control of several insect pests, acting as both antifeedants and pesticides. (G. Shay)

Palms

Several salt-tolerant palms are sources of fiber and other materials for a wide variety of uses. Fronds from the nipa palm (see p. 66), for example, are also used for thatching, basketry, mats, and similar applications.

Cocos nucifera, the familiar coconut palm, is commonly found on sandy beaches but also occurs in low marshy areas occasionally flooded by seawater. The uses of the nut for food, its value as a

Many attractive halophytes can be used as landscape plants or for floral use, especially in areas with constraints on the use of fresh water. Sea lavender, for example, can be irrigated with seawater and used to produce cut flowers. (G. Shay)

source of oil, and tapping of the inflorescence for toddy and sugar are all well known. In addition, leaves are used for thatching, walls, and screens, and leaflets are woven into baskets, plates, hats, mats, and other articles for daily use. Fibers from the husk are used for brushes, mats, twine, rope, stuffing for mattresses and upholstery, and for caulking boats.

Elaeis oleifera is found in coastal swamp forests from the lower basin of the Amazon to southern Mexico. Often called the American

A potential ornamental, the low-growing succulent *Arthrocnemum fruticosum* tolerates seawater irrigation in field trials in Israel. (J.A. Aronson)

oil palm, it is closely related to the African oil palm (*E. guineensis*). The fruits are a source of oil, tallow, and chicken feed. The tree has a low-growing habit, which simplifies fruit harvest.

Licuala spinosa is a palm found in tidal forests immediately behind the mangroves from the Malay Peninsula to the Andaman Islands. Its leaves are used for roofing and for wrapping food.

Manicaria saccifera, the monkey-cap palm, occurs in tidal swamps in Central America and northern South America. The leaves are used for roofing and for covering boats. The fibrous inner bract is made into fiber for bags, mats, hats, and other personal articles. Its oil is suitable for soapmaking.

Oncosperma filimentosa, the nibung palm, grows on brackish lowlands just behind mangrove stands in India, Sri Lanka, and the Philippines. The trunk is used for construction and the spines are used as darts in blowpipes and as tips on fish spears. The leaves are used for making baskets and the bracts as buckets.

Raphia taedigera, the pine cone palm, is found in marshes from the lower Amazon north to Costa Rica. Its multiple short trunks are

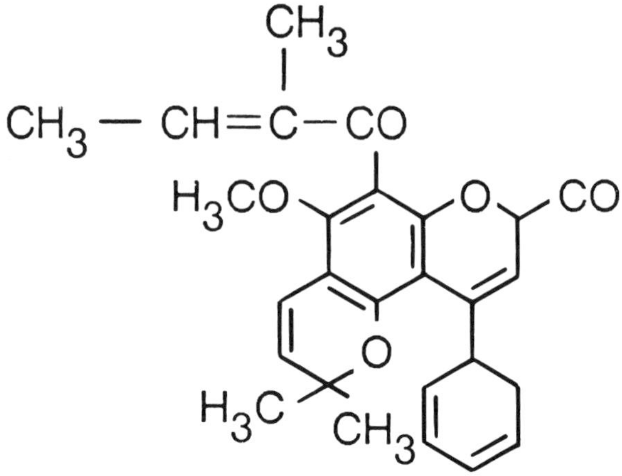

FIGURE 4 Callophyllolide. A promising agent for the treatment of inflammatory and rheumatic conditions, callophyllolide, has been isolated from the seeds of *Calophyllum inophyllum*, a plant of coastal southern India, Burma, and Sri Lanka. Source: R. C. Saxena et al., 1982

used for walls in native dwellings and its fibers are used for fishing nets and cordage.

Raphia vinifera, the bamboo palm, grows in tidal bays and creeks in tropical West Africa. Fiber from the leaf bases is used for fishing lines and for animal snares and cordage. It is also exported for use in the manufacture of brooms, industrial brushes, and upholstery stuffing.

BIOACTIVE DERIVATIVES

Callophyllolide (Figure 4), a complex 4-phenyl coumarin, has been isolated from the seeds of *Calophyllum inophyllum* (Alexandrian laurel), a common evergreen tree of coastal southern India, Burma, and Sri Lanka. In preliminary testing against oxyphenbutazone, a widely prescribed antiinflammatory, callophyllolide appears promising for the treatment of inflammatory and rheumatic conditions. The seed oil can also be saponified to give a soap with antibacterial properties. Extracts of the bark and leaves of *C. inophyllum* are used in traditional Indian medicine. These extracts have been qualitatively analyzed to show the presence of steroids and alkaloids.

Another potential ornamental, *Nitraria billardieri*, has voluminous growth with seawater irrigation. (G. Shay)

The fruits of *Balanites roxburghii* are a potential source of diosgenin, a precursor for the synthesis of a number of steroidal drugs. The ripe fruits can contain 3-4 percent diosgenin. Diosgenin has great versatility. Progesterone and orally active progesterone analogues can be readily produced from diosgenin through chemical synthesis and cortisone and other useful corticosteroid hormones by microbial synthesis. For the manufacture of most current contraceptive drugs, however, the preferred starting material is sitosterol from soybean oil extraction wastes.

A recent discussion paper published by FAO suggests that diosgenin could be produced in the Sudan from the fruits of *Balanites aegyptiaca*. The paper calculates that the Sudan could produce 1,200 tons per year, enough to meet about half the world demand and earn an export income of $36 million.

The neem tree (*Azadirachta indica*) has great potential for agricultural and commercial exploitation. It is a fast-growing tree that can be established on poor soils unsuitable for farming. Fruiting begins at about five years. Although neem seed oil is inedible, it is traditionally used in soapmaking. More importantly, neem seed extracts are effective in the control of several insect pests. The extracts act both as antifeedants and pesticides, and appear to be nontoxic to humans and animals. Neem seedlings have been grown successfully in Pakistan on sandy soil using irrigation water with an EC of 17 dS/m. A neem plantation has been established near Mecca in Saudi Arabia to provide shade for Muslim pilgrims. Water with an EC of 4.25 dS/m is used for irrigation.

Various extracts of *Adhatoda vasica*, a salt-tolerant evergreen shrub common in India, are effective as antifeedants, insecticides, viricides, and as wound healing agents for water buffalo. The alkaloid vasicine is found in the leaves and bark at 0.2-0.4 percent. Extracts are also used in commercial preparations for the treatment of asthma and bronchitis. *A. vasica* has also been grown as a firewood crop. (See p. 63).

The perennial herb *Anemopsis californica* is found in semisaline and alkaline wetland soils in the southwestern United States and northwestern Mexico. It has long been esteemed for medicinal purposes in this region and continues to be widely used in Sonora. Extracts from the large roots (up to a meter in length) are used internally for colds, coughs, or indigestion, and externally for wounds or swellings. One of the active ingredients appears to be 4-allylveratrole, a mild antispasmodic.

Periwinkle (*Catharanthus roseus*), a tropical plant found in the coastal sands of India, has been found to grow under saline conditions (up to 12 dS/m). Catharanthus roots contain alkaloids used in the treatment of leukemia, and its leaves contain alkaloids reported to lower blood pressure.

TABLE 16 Bioactive Materials Isolated From *Salsola* Species.

Species	Compound	Amount	Potential Use
S. richteri	Salsolinol	---	Heart stimulant, binds to brain neuroreceptors
S. kali S. richteri S. ruthenica	Salsolidine and Salsoline	0.2% 0.2%	Antihypertensives, vasodilators
S. subaphylla	N-feruloyl-putrecine	---	Weakly antihypertensive
S. pestifer	Carotene	4.5 mg/100g	Vitamin
S. pestifer	Ascorbic acid	77 mg/100g	Vitamin
S. gemmascens	Citric acid	3.4%	Food additive

SOURCE: Adapted from Fowler, 1985.

Derris trifoliata, a climbing vine, abounds in mangrove forests and along muddy shores from East Africa to India and through Malaysia to Polynesia. The leaves, which contain rotenone, are pounded and put in shallow water to stun fish. Such toxicants can be used to eliminate predators and competitors in freshwater and brackish water ponds to be used for the culture of crustaceans and finfish.

Other mangrove vegetation has similar uses. The bark and seeds of *Aegiceras corniculatum*, *Avicennia alba*, *Barringtonia asiatica*, and the roots of *Heritiera littoralis* all contain a fish poison, and the milky sap of *Excoecaria agallocha* is used as a fish and arrowhead poison.

Citrullus colocynthis, a creeping plant, occurs in the warmer parts of Asia and Africa. It is common on the seashores of western India and is used to control drifting sand in coastal Pakistan. The dried pulp of its unripe, full-grown fruit constitutes the drug colocynth, which is used as a cathartic.

In addition to the potential for Russian-thistle (*Salsola iberica*) as a fodder source (p. 76), other *Salsola* species contain recoverable amounts of bioactive materials. Some of these are shown in Table 16.

TABLE 17 Salt-Tolerant Ornamental Plants.

Plant	Flower Color	Flowering Season	Average Height(m)	Salt Resistance
Trees:				
Acacia gerrardii	cream	July-Oct	5	1
A. horrida	yellow	May-Sept	8	1
A. raddiana	cream	Mar-Apr Oct-Dec	5	1
A. salicina	cream	Mar and Sept	8	1
A. tortilis	cream	Spring and Fall	5	1
Casuarina glauca		Apr	8	1
Conocarpus erectus			5	1
Elaeagnus angustifolia	white	Apr-May	4	1
Eucalyptus sargentii	yellowish-white		6	2
Moringa peregrina	white to pink	Mar-May	6	1
Parkinsonia aculeata	yellow	May-June	8	2
Phoenix dactylifera			12	2
Prosopis juliflora	white	Apr-May	6	1
Tamarix aphylla	white	May-June	8	2
Shrubs:				
Atriplex barclayana			1.5	2
A. cinerea			1	2
A. nummularia			1.5-2	2
Callistemon rigidus	red	Apr	2	1
Cassia mexicana	yellow	Apr-Sept	0.5-0.75	1
Colutea istria	yellow	Mar-Apr	2	1
Maireana sedifolia			2	2
Melaleuca nesophila	lilac	May-July	3	1
Retama raetam	white/purple	Mar-Apr	2	1
Tamarix chinensis "mapu"	violet		3	1

LANDSCAPE AND ORNAMENTAL USE

Many attractive halophytes can be used as landscape plants, especially in areas with constraints on the use of fresh water for watering or irrigation. In Israel, trees such as *Conocarpus erectus, Eucalyptus sargentii,* and *Melaleuca halmaturorum,* and shrubs such as *Maireana sedifolia, Borrichea frutescens,* and *Clerodendrum inerme* are sold for amenity planting to allow irrigation with saline

TABLE 17 (Continued)

Plant	Flower Color	Flowering Season	Average Height(m)	Salt Resistance
Succulents and Semi-Succulents:				
Agave americana	white	Apr-July	2	1
*Arthrocnemum fruticosum**			0.6	3
A. macrostachyum			0.5	3
Batis maritima			0.3	3
Biennial and Perennial Ground Cover:				
Arctotis grandis	assorted	Dec-Apr		1
Aster alpinus	blue	Dec-Apr		1
Catharansus roseus	white/pink	Most of year		1
Cineraria maritima	violet	Apr-June		1
Crithmum maritimum	yellow	May-June		2
Gazania splendens	assorted	Dec-May		1
Inula crithmoides	yellow	June-July		3
Nitraria billardieri	white	Apr-May		3
Sesuvium verrucosum	lilac	June-July		3
Lawn Grasses:				
Cynodon dactylon				2
Paspalum vaginatum				2

Salt resistance: arbitrary degrees according to soil electrical conductivity:
1 = 5-15 dS/m; 2 = 15-25 dS/m; 3 = 25-50 dS/m.

*Thrives under conditions of waterlogging.

SOURCE: Adapted from Pasternak et al., 1986.

water. A selection of salt-tolerant ornamental plants is shown in Table 17. The striking floral display of the *Butea monosperma* tree (p. 63) has earned it the name "flame of the forest." In addition, plants such as *Limonium* species have potential for floral use. For example, sea lavender (*Limonium axillare*) can be irrigated with seawater and used to produce cut flowers.

REFERENCES AND SELECTED READINGS

General

Balandrin, M. F., J. A. Klocke, E. S. Wurtele and W. H. Bollinger. 1985. Natural plant chemicals: sources of industrial and medicinal materials. *Science* 228:1154-1160.
Hinman, C. W. 1984. New crops for arid lands. *Science* 225:1445-1448.
Vietmeyer, N. D. 1986. Lesser-known plants of potential use in agriculture and forestry. *Science* 232:1379-1384.

Essential Oils

Kewda

Dutta, P. K., H. O. Saxena and M. Brahman. 1987. Kewda perfume industry in India. *Economic Botany* 41:403-410.

Mentha and Other Species

Chandra, V., A. Singh and L. D. Kapoor. 1968. Experimental cultivation of some essential oil bearing plants in saline soils. *Perfume and Essential Oil Review*. December:869-873.
Graven, E. H., B. Gardner and C. Tutt. 1987. Essential oils—new crops for Southern Africa. *Ciskei Agricultural Journal* 1:2-8.
Patra, P. and P. K. Dutta. 1979. Studies on salinity tolerance in aromatic and medicinal plants. *Journal of the Orissa Botanical Society* 1(1):17-18.
Piprek, S. R. K., E. H. Graven and P. Whitfield. 1982. Some potentially important indigenous aromatic plants for the eastern seaboard areas of Southern Africa. *World Crops* 10(4):255-263.

Gums, Oils and Resins

General

Forti, M. 1986. Salt tolerant and halophytic plants in Israel. *Reclamation and Revegetation Research* 5:83-96.
Greek, B. F. 1987. Modest growth ahead for rubber. *Chemical and Engineering News* 66(12):25-51.

Sesbania

Chahda, Y. R. (ed.). 1972. *Sesbania. The Wealth of India*. IX:293-303. CSIR, New Delhi, India.
Chandra, V. and M. I. H. Farooqi. 1979. *Dhaincha* for seed gum. *Extension Bulletin No. 1*. National Botanical Research Institute, Lucknow, India.

Gorham, J., E. McDonnell and R. G. Wyn Jones. 1984. Pinitol and other solutes in salt-stressed *Sesbania aculeata*. *Zeitschrift fur Pflanzenphysiologie* 114:173-178.

Grindelia

Hoffmann, J. J. and S. P. McLaughlin. 1986. *Grindelia camporum*: potential cash crop for the arid southwest. *Economic Botany* 40:162-169.

Schuck, S. M. and S. P. McLaughlin. 1988. Flowering phenology and outcrossing in tetraploid *Grindelia camporum* Greene. *Desert Plants* 9(1):7-16.

Timmerman, B. N. and J. J. Hoffmann. 1985. The potential for the commercial utilization of resins from *Grindelia camporum*. Pp. 1321-1339. in: E. E. Whitehead, C. F. Hutchinson, B. N. Timmermann, and R. G. Varady (eds.) *Arid Lands Today and Tomorrow*. Westview Press, Boulder, Colorado, US.

Larrea tridentata

Belmares, H. and A. Barrera. 1979. Polymerization studies of creosote bush (*Larrea tridentata*) phenolic resin with formaldehyde. *Journal of Applied Polymer Science* 24:1531-1537.

Belmares, H., A. Barrera, M. Ortega and M. Monjaras. 1980. Adhesives from creosote bush (*Larrea tridentata*) phenolic resin with formaldehyde. Characteristics and application. *Journal of Applied Polymer Science* 25:2115-2118.

Sapium sebiferum

Chadha, Y. R. (ed.). 1972. *Sapium. Wealth of India*. IX:229-231. CSIR, New Delhi, India.

Scheld, H. W. and J. R. Cowles. 1981. Woody biomass potential of the Chinese tallow tree. *Economic Botany* 35:391-397.

Scheld, H. W., J. R. Cowles, C. R. Engler, R. Kleiman and E. B. Shultz, Jr. 1984. Seeds of the Chinese tallow tree as a source of chemicals and fuels. Pp. 81-101 in: E. B. Shultz, Jr. and R. P. Morgan (eds.) *Fuels and Chemicals from Oilseeds*. Westview Press, Boulder, Colorado, US.

Jojoba

Baldwin, A. R. (ed.) 1988. *Proceedings: 7th International Conference on Jojoba and its Uses*. American Oil Chemists Society, Chicago, Illinois, US.

Bhatia, V. K., A. Chaudhry, A. Masohan, R. P. S. Bisht and G. A. Sivasankaran. 1988. Sulphurization of jojoba oil for application as extreme pressure additive. *Journal of the American Oil Chemists Society* 65(9):1502-1507.

Guayule

François, L. E. 1986. Salinity effects on four arid zone plants (*Parthenium argentatum, Simmondsia chinensis, Kochia prostrata* and *Kochia brevifolia*). *Journal of Arid Environments* 11:103-109.

Hoffman, G. J., M. C. Shannon, E. V. Maas, L. Grass. 1988. Rubber production of salt-stressed guayule at various plant populations. *Irrigation Science* 9:213-226.

Maas, E. V., T. J. Donovan and L. E. François. 1988. Salt tolerance of irrigated guayule. *Irrigation Science* 9:199-212.

Miyamoto, S. and D. A. Bucks. 1985. Water quantity and quality requirements of guayule: current assessment. *Agricultural Water Management* 10:205-219.

Chrysothamnus

Ostler, W. K., C. M. McKell and S. White. 1986. *Chrysothanus nauseosus*: a potential source of natural rubber. Pp. 389-394 in *Proceedings - Symposium on the Biology of Artemisia and Chrysothamnus*. USDA, Ogden, Utah, US.

Weber, D. J., D. F. Hegerhorst, T. D. Davis and E. D. McArthur. 1987. Potential uses of rubber rabbitbrush (*Chrysothamnus nauseosus*). Pp. 27-33 in: K. L. Johnson (ed.) *The Genus Chrysothamnus*. Utah State University, Logan, Utah, US.

Pulp and Fiber

Reed

de la Cruz, A. A. 1978. The production of pulp from marsh grass. *Economic Botany* 32:46-50.

de la Cruz, A. A. and G. R. Lightsey. 1981. *Pulping Characteristics and Paper Making Potential of Non-wood Wetland Plants*. Sea Grant Publication MASGP-80-016. Ocean Springs, Mississippi 39564, US.

Graneli, W. 1984. Reed *Phragmites australis* as an energy source in Sweden. *Biomass* 4:183-208.

Iyengar, E. R. R. and J. B. Pandya. 1983. *Juncus rigidus* for saline soils. *Indian Journal of Agricultural Chemistry* 16(1):147-152.

Zahran, M. A. and M. A. El Demerdash. 1984. Transplantation of *Juncus rigidus* in the saline and non-productive lands of Egypt. Pp. 75-131 *Research in Arid Zones*. Report No. 17, International Foundation for Science, Stockholm, Sweden.

Zahran, M. A. 1986. Establishment of fiber producing halophytes in salt affected areas of Egypt. Pp. 235-251 in: R. Ahmad and A. San Pietro (eds.) *Prospects for Biosaline Research*. University of Karachi, Karachi, Pakistan.

Typha

Morton, J. F. 1975. Cattails (*Typha* spp.) - a weed problem or potential crop? *Economic Botany* 29:7-29.

Saccharum griffithii

Ahmad, R. 1987. *Saline Agriculture at Coastal Sandy Belt*. University of Karachi, Karachi, Pakistan.

Hibiscus

François, L. E., T. J. Donovan and E. V. Maas. 1988. Salt tolerance of kenaf. Presented at 1st National Symposium for New Crops: Research, Development, Economics. October 23-26, 1988. Indianapolis, Indiana, US.

Kugler, D. E. 1988. *Kenaf Newsprint: Realizing Commercialization After Four Decades of Research and Development.* USDA, Washington, DC, US.

Sastri, B. N. (ed.). 1959. *Hibiscus. The Wealth of India.* V:75-98. CSIR, New Delhi, India.

Cotton

Ayars, J. E., R. B. Hutmacher, R. A. Schoneman, S. S. Vail and D. Felleke. 1986. Drip irrigation of cotton with saline drainage water. *Transactions of the ASAE* 29(6):1668-1673.

Babu, V. R., S. N. Prasad, A. M. Babu and D. S. K. Rao. 1987. Evaluation of cotton genotypes for tolerance to saline water irrigations. *Indian Journal of Agronomy* 32(3):229-231.

Bouzaidi, A. and S. El Amami. 1980. Irrigation a l'eau salee de deux varietes de cotonnier dans les essais de plein champ. *Physiologie Vegetale* 18(1):35-44.

Dean, P. 1981. Two-bale cotton with high-salt water. *Agricultural Research* (October):10-11.

Iyengar, E. R. R., J. B. Pandya and J. S. Patolia. 1978. Evaluation of cotton varieties to salinity stress. *Indian Journal of Plant Physiology* 21(2):113-117.

Mantell, A., H. Frenkel and A. Meiri. 1985. Drip irrigation of cotton with saline-sodic water. *Irrigation Science* 6:95-106.

Nawaz, A., N. Ahmad and R. H. Qureshi. 1986. Salt tolerance of cotton. Pp. 285-291 in: R. Ahmad and A. San Pietro (eds.) *Prospects for Biosaline Research.* University of Karachi, Karachi, Pakistan.

Palms

Balick, M. J. 1979. Amazonian oil palms of promise: A survey. *Economic Botany* 33:11-28.

Morton, J. F. 1976. Craft industries from coastal wetland vegetation. Pp. 254-266 in: M. Wiley (ed.) *Estuarine Processes.* Vol. 1. Academic Press, New York, New York, US.

Pinheiro, C. U. B. and M. J. Balick. 1987. *Brazilian Palms: Notes on their Uses and Vernacular Names.* New York Botanical Garden, Bronx, New York, US.

Plotkin, M. J. and M. J. Balick. 1984. Medicinal uses of South American palms. *Journal of Ethnopharmacology* 10(2):157-179.

Bioactive Derivatives

Calophyllum inophyllum

Mehrotra, S., R. Mitra and H. P. Sharma. 1986. Pharmacognostic studies on punnaga, *Calophyllum inophyllum* L., leaf and stem bark. *Herba Hungaria* 25(1):45-71.

Saxena, R. C., R. Nath, G. Palit, S. K. Nigam and K. P. Bhargava. 1982. Effect of calophyllolide, a nonsteroidal anti-inflammatory agent, on capillary permeability. *Planta Medica: Journal of Medicinal Plant Research* 44(4):246-248.

Guevara, B. Q. and R. C. Solevilla. 1983. An antibacterial soap from bitaog oil. *Acta Manilana A, Natural and Applied Sciences* 22(11):62-64.

Balanites roxburghii

Ghanim, A., I. Chandrasekharan, V. A. Amalraj and H. A. Khan. 1984. Studies on diosgenin content in fruits of *Balanites roxburghii*. *Transactions of the Indian Society of Desert Technology and University Center of Desert Studies* 9(2):21-22.

National Research Council. 1987. *Workshop on Biotechnology of Steroid Compounds as Contraceptives and Drugs. Summary Report.* National Research Council, Jakarta, Indonesia, and National Academy Press, Washington, DC, US.

Azadirachta indica

Ahmed, S., S. Bamofleh and M. Munshi. 1989. Cultivation of neem (*Azadirachta indica*, Meliaceae) in Saudi Arabia. *Economic Botany* 43:35-38.

Ahmed, S. and M. Grainge. 1986. Potential of the neem tree (*Azadirachta indica*) for pest control and rural development. *Economic Botany* 40:201-209.

Deshmukh, P. B. and D. M. Renapurkar. 1987. Insect growth regulatory activity of some indigenous plant extracts. *Insect Science and Its Application* 8(1):81-83.

Kazmi, S. M. A. 1980. *Melia azadirachta*—A most common cultivated tree in Somalia. *Somalia Range Bulletin* 9:20-23.

Radwanski, S. A. and G. E. Wickens. 1981. Vegetative fallows and potential value of the neem tree (*Azadirachta indica*) in the tropics. *Economic Botany* 35:398-414.

Saxena, R. C. 1989. Insecticides from neem. Pp. 110-135 in: J. T. Arnason, B. J. R. Philogene and P. Morand (eds.) *Insecticides of Plant Origin*. American Chemical Society, Washington, DC, US.

Adhatoda vasica

Arambewela, L. S. R., C. K. Ratnayake, J. S. Jayasekera and K. T. D. De Silva. 1988. Vasicine contents and their seasonal variation in *Adhatoda vasica*. *Fitoterapia* 59(2):151-153.

Bhargava, M. K., H. Singh, A. Kumar and K. C. Varshney. 1986. *Adhatoda vasica* as wound healing agent in buffaloes - histological and histochemical studies. *Indian Journal of Veterinary Surgery* 7(2):29-35.

Saxena, B. P., K. Tikku, C. K. Atal and O. Koul. 1986. Insect antifertility and antifeedant allelochemics in *Adhatoda vasica*. *Insect Science and Its Application* 7(4):489-493.

Tripathi, R. N., R. K. R. Tripathi and D. K. Pandey. 1981. Assay of antiviral activity in the crude leaf sap of some plants. *Environment India* 4(1/2):86-87.

Anemopsis californica

Childs, R. F. and J. R. Cole. 1965. Phytochemical and pharmacological investigation of *Anemopsis californica*. *Journal of Pharmaceutical Sciences* 54(5):789-791.

Ezcurra, E., R. S. Felger, A. D. Russell and M. Equihua. 1988. Freshwater islands in a desert sand sea: the hydrology, flora, and phytogeography of the Gran Desierto oases of northwestern Mexico. *Desert Plants* 9(2):35-44,55-63.

Mangrove Toxicants

De la Cruz, A. A., E. D. Gomez, D. H. Miles, G. J. B. Cajipe and V. B. Chavez. 1984. Toxicants from Mangrove plants: I. Bioassay of crude extracts. *International Journal of Ecological and Environmental Science* 10:1-9.

Gomez, E. D., A. A. de la Cruz, V. B. Chavez, D. H. Miles and G. J. B. Cajibe. 1986. Toxicants from mangrove plants: II. Toxicity of aqueous extracts to fish. *The Philippine Journal of Science* 115(2):81-89.

Miles, D. H., D.-S. Lho, A. A. de la Cruz, E. D. Gomez, J. A. Weeks and J. L. Atwood. 1987. Toxicants from mangrove plants III. Heritol, a novel ichthyotoxin from the mangrove plant *Heritiera littoralis. Journal of Organic Chemistry* 52:2930-2932.

Citrullus colocynthis

Bringi, N. V. 1987. Lesser known tree-borne oil seeds. Pp. 216-248 in: N. V. Bringi (ed.) *Non-Traditional Oils and Oilseeds of India.* Oxford and IBH Publishing Co., New Delhi, India.

Sastri, B. N. (ed.). 1950. *Citrullus colocynthis. Wealth of India.* II:185-186. CSIR, New Delhi, India.

Russian-Thistle

Fowler, J. L., J. H. Hageman, and M. Suzukida. 1985. *Evaluation of the Salinity Tolerance of Russian Thistle to Determine its Potential for Forage Production Using Saline Irrigation Water.* New Mexico Water Resources Institute, Las Cruces, New Mexico, US.

Landscape and Ornamental Use

Pasternak, D., J. A. Aronson, J. Ben-Dov, M. Forti, S. Mendlinger, A. Nerd and D. Sitton. 1986. Development of new arid zone crops for the Negev Desert of Israel. *Journal of Arid Environments* 11:37-59.

Verkade, S. D. and G. E. Fitzpatrick. 1988. Development of the threatened halophyte *Mallontonia gnaphalodes* as a new ornamental crop. Presented at 1st National Symposium for New Crops: Research, Development, Economics. October 23-26, 1988. Indianapolis, Indiana, US.

RESEARCH CONTACTS

Essential Oils

V. Chandra, National Botanic Gardens, Lucknow 226001, India

P. K. Dutta, Aromatic and Medicinal Plants Division, Regional Research Laboratory, Bhubaneswar 751 013, Orissa, India.

Earl Graven, Head, Department of Agronomy, University of Fort Hare, Alice 5700, Ciskei, South Africa.

Gums, Oils and Resins

General

V. Chandra, National Botanic Gardens, Lucknow 226001, India.
M. Forti, The Institutes for Applied Research, Ben Gurion University, PO Box 1025, 84110 Beer-Sheva, Israel.
E. Rodriguez, Phytochemical Laboratory, School of Biological Sciences, University of California, Irvine CA 92717, US.

Dhaincha

R. G. Wyn Jones, Center for Arid Zone Studies, University College of North Wales, Bangor, Wales, LL57 2UW, UK.
V. Chandra, National Botanical Research Institute, Lucknow 226001, India.

Grindelia

Stephen P. McLaughlin, Bioresources Research Facility, University of Arizona, 250 E. Valencia Road, Tucson, AZ 85706, US.

Creosote Bush

Hector Belmares, Centro de Investigascion en Quimica Aplicada, Aldama Ote. 371, Saltillo, Coahuila, Mexico.

Chinese Tallow Tree

Robert Kleiman, USDA Northern Regional Research Center, 1815 North University, Peoria, IL 61604, US.
H. W. Scheld, PhytoResource Research, Inc., 707 Texas Avenue - Suite 202D, College Station, TX 77840, US.
E. B. Shultz, Jr., Box 1106, Washington University, St. Louis, MO 63130, US.

Jojoba

Hal C. Purcell, Jojoba Grower's Association, 3420 East Shea - Suite 125, Phoenix, AZ 85028, US.
David Palzkill, Department of Plant Sciences, University of Arizona, Tucson, AZ 85721, US.
John Rothfus, USDA Northern Regional Research Center, 1815 North University, Peoria, IL 61604, US.

Chrysothamnus nauseosus

D. J. Weber, Department of Botany and Range Science, Brigham Young University, Provo, UT 84602, US.

Guayule

Joseph Beckman, Firestone Tire and Rubber Company, 1200 Firestone Parkway, Akron, OH 44317, US.
D. A. Bucks, US Water Conservation Laboratory, 4331 East Broadway, Phoenix, AZ 85040, US.
S. Miyamoto, Texas Agricultural Experiment Station, 1380 A&M Circle, El Paso, TX 79927, US.

Pulp and Fiber

Cotton

James E. Ayars, USDA-ARS, WMRL, Fresno, CA 93700, US.
A. Mantell, Institute of Soils and Water, Agricultural Research Organization, The Volcani Center, Bet Dagan, Israel.
Akhtar Nawaz, Department of Soil Science, University of Agriculture, Faisalabad, Pakistan.
James D. Rhoades, USDA Salinity Research Laboratory, 4500 Glenwood Drive, Riverside, CA 92501, US.

Reed

Armando A. de la Cruz, Department of Zoology, Mississippi State University, MI 39762, US.
Wilhelm Graneli, Institute of Limnology, Box 3060, S-22003 Lund, Sweden.
E. R. R. Iyengar, Central Salt and Marine Chemicals Research Institute, Bhavnagar 364 002, India.
M. A. Zahran, Botany Department, Mansoura University, Mansoura, Egypt.

Esparto Grass

Director, Institut National de Recherches Forestieres, Ministere de l'Agriculture, Route de la Soukra, B.P. 2, Ariana, Tunisia.

Typha

J. F. Morton, Morton Collectanea, University of Miami, Coral Gables, FL 33124, US.

Kenaf/Hibiscus

Charles Adamson, USDA Plant Introduction Center, Route 1, Sharpsburg, GA 30277, US.
Marvin O. Bagby, USDA Northern Regional Research Center, 1815 North University, Peoria, IL 61604, US.
L. E. François, US Salinity Laboratory, 4500 Glenwood Drive, Riverside, CA 92501, US.

Palms

M. J. Balick, New York Botanical Garden, Bronx, NY 10458, US.
J. F. Morton, Morton Collectanea, University of Miami, Coral Gables, FL 33124, US.

Bioactive Derivatives

Calophyllum inophyllum

S. Mehrotra, Pharmacognosy Section, National Botanical Research Institute, Lucknow 22600, India.
R. C. Saxena, Department of Pharmacology, King George's Medical College, Lucknow 226003, India.

Balanites roxburghii

A. Ghanim, Central Arid Zone Research Institute, Jodhpur 342 003, India.

Azadirachta indica

S. A. Radwanski, Land Resources Consultancy, 361 Wimbledon Park Road, London SW19 6PE, UK.
G. E. Wickens, Royal Botanic Gardens, Kew, Surrey, TW9 3AB, UK.

Commiphora wightii

S. Kumar and V. Shankar, Central Arid Zone Research Institute, Jodhpur 342003, India.

Catharanus roseus

P. K. Dutta, Aromatic and Medicinal Plants Division, Regional Research Laboratory, Bhubaneswar 751 013, Orissa, India.

Landscape and Ornamental Use

Dov Pasternak, Institute for Desert Research, Ben Gurion University, Sede Boger 84990, Israel.
Stephen D. Verkade, University of Florida, 3205 College Avenue, Fort Lauderdale, FL 33314, US.

Index

A

Acacia species, 23, 62, 92
Adhatoda vasica, 63, 118
Alexandrian Laurel, 116
Alkali Sacaton, 20, 81
American Oil Palm, 114
Anemopsis californica, 118
Argan, 25
Argania spinosa, 25
Arthrocnemum fruticosum, 115
Asparagus, 35
Asparagus officinalis, 35
Atriplex triangularis, 27
Atriplex species, 27, 31, 81, 83
Azadirachta indica, 118

B

Bajra, 20
Balanites species, 117
Bamboo Palm, 115
Barley, 36
Batis maritima, 26
Beta vulgaris, 17, 66
Butea monosperma, 63, 121

C

Calophyllum inophyllum, 116
Casuarina species, 56
Catharanthus roseus, 118
Channel Millet, 78
Chenopodium quinoa, 20
Chinese Tallow Tree, 105
Chloris gayana, 80
Chrysothamnus nauseosus, 108
Citrullus colocynthis, 119
Coccoloba uvifera, 33
Coconut Palm, 113
Cocos nucifera, 113
Common Indian Saltwort, 28
Common Purslane, 26
Common Reed, 109
Cordgrasses, 79
Cotton, 112
Creosote Bush, 105
Crithmum maritimum, 27

D

Derris trifoliata, 119
Dhaincha, 105
Distichlis palmeri, 19
Distichlis species, 77

E

Echinochloa turnerana, 78
Eelgrass, 19
Elaeis oleifera, 114
Eleocharis dulcis, 26
Elytrigia elongatum, 36, 81
Esparto Grass, 110
Eucalyptus species, 55

G

Gossypium hirusutum, 112
Grindelia species, 105
Guayule, 108

H

Halosarcia species, 91
Hedysarum carnosum, 81
Hibiscus cannabinus, 111
Hibiscus tiliaceus, 111
Honey, 38
Hordeum vulgare, 36

I

Ice Plant, 27
Indian Almond, 23

J

Jojoba, 107
Juncus species, 109

K

Kallar Grass, 67, 75
Karanjin, 63
Kenaf, 111
Kewda, 103
Kochia species, 31, 90
Kosteletzkya virginica, 21

L

Larrea tridentata, 105
Leptochloa fusca, 67, 75
Leuceana leucocephala, 92
Licuala spinosa, 115
Limonium species, 121
Lycium fremontii, 33

M

Mairiena species, 81, 89
Maize, 38
Mangroves, 58, 119
Manicaria saccifera, 115
Manila Tamarind, 63
Melaleuca species, 60
Mentha species, 104
Mesembryanthemum crystallinum, 27

Monkey Cap Palm, 115

N

Neem, 118
Nibung Palm, 115
Nipa Palm, 66, 113
Nitraria billardieri, 117

O

Oncosperma filimentosa, 115
Oryza sativa, 36

P

Pandanus fascicularis, 103
Pandanus tectoris, 110
Palmer Saltgrass, 19
Parthenium argentatum, 108
Paspalum vaginatum, 75
Pearl Millet, 20
Pennisetum typhoides, 20
Periwinkle, 118
Phragmites australis, 109
Pine Cone Palm, 115
Pithecellobium dulce, 63
Pongamia pinnata, 63
Portulaca oleracea, 26
Prosopis species, 52, 93
Puccinellia species, 81

Q

Quandong, 33
Quinoa, 20

R

Raphia taedigera, 115
Raphia vinifera, 115
Rhizophora species, 58
Rhodes Grass, 80
Rice, 36
Rubber Rabbitbrush, 108
Rush, 109
Russian-Thistle, 76, 119

S

Saccharum griffithii, 111
Salicornia species, 25, 31

Salsola iberica, 76, 119
Salsola species, 31, 119
Saltgrasses, 77
Saltwort, 26
Salvadora species, 32
Samphire, 91
Sand Couch, 38
Santalum acuminatum, 33
Sapium sebiferum, 105
Sea Fennel, 27
Seagrape, 33
Seashore Mallow, 21
Seaside Purslane, 26
Sesbania bispinosa, 105
Sesuvium portulacastrum, 26
Silt Grass, 75
Simmondsia chinensis, 107
Southern Cattail, 110
Spartina species, 79
Sporobolus airoides, 20, 81
Sporobolus species, 20, 81
Stipa tenacissima, 110
Suaeda maritima, 28

T

Tall Wheatgrass, 36, 81
Tamarix species, 61
Tecticornia species, 23
Terminalia catappa, 23
Textile Screwpine, 110
Thinopyrum bessarabicum, 38
Triticum aestivum, 36
Typha domingensis, 110

U

Urochondra setulosa, 112

W

Wheat, 36
Wild Water Chestnut, 26

Z

Zea mays, 38
Zostera marina, 19

Board on Science and Technology for International Development

RALPH H. SMUCKLER, Dean of International Studies and Programs, Michigan State University, East Lansing, *Chairman*

Members

JORDON J. BARUCH, President, Jordan Baruch Associates, Washington, D.C.

PETER D. BELL, President, The Edna McConnell Clark Foundation, New York, New York

GEORGE T. CURLIN, The Fogarty International Center, The National Institutes of Health, Bethesda, Maryland

DIRK FRANKENBERG, Director, Marine Science Program, University of North Carolina, Chapel Hill

ELLEN L. FROST, Corporate Director, International Affairs, United Technologies Corporation, Washington, D.C.

FREDERICK HORNE, Dean of the College of Science, Oregon State University, Corvallis

ROBERT KATES, Director, Alan Shaw Feinstein World Hunger Program, Brown University, Providence, Rhode Island

CHARLES C. MUSCOPLAT, Executive Vice President, Molecular Genetics, Inc., Minnetonka, Minnesota

ANTHONY SAN PIETRO, Professor of Plant Biochemistry, Indiana University, Bloomington

ALEXANDER SHAKOW, Director, Department of Strategic Planning and Review, The World Bank, Washington, D.C.

BARBARA D. WEBSTER, Associate Dean, Office of Research, University of California, Davis

GERALD P. DINEEN, Foreign Secretary, National Academy of Engineering, *ex officio*

WILLIAM E. GORDON, Foreign Secretary, National Academy of Sciences, *ex officio*

Board on Science and Technology for International Development
Publications and Information Services (HA-476E)
Office of International Affairs
National Research Council
2101 Constitution Avenue, N.W.
Washington, D.C. 20418 USA

How to Order BOSTID Reports

BOSTID manages programs with developing countries on behalf of the U.S. National Research Council. Reports published by BOSTID are sponsored in most instances by the U.S. Agency for International Development. They are intended for distribution to readers in developing countries who are affiliated with governmental, educational, or research institutions and who have professional interest in the subject areas treated by the reports.

BOSTID books are available from selected international distributors. For more efficient and expedient service, please place your order with your local distributor. (See list on back page.) Requestors from areas not yet represented by a distributor should send their orders directly to BOSTID at the above address.

Energy

33. **Alcohol Fuels: Options for Developing Countries.** 1983, 128pp. Examines the potential for the production and utilization of alcohol fuels in developing countries. Includes information on various tropical crops and their conversion to alcohols through both traditional and novel processes. ISBN 0-309-04160-0.

36. **Producer Gas: Another Fuel for Motor Transport.** 1983, 112pp. During World War II Europe and Asia used wood, charcoal, and coal to fuel more than a million gasoline and diesel vehicles. However, the technology has since been virtually forgotten. This report reviews producer gas and its modern potential. ISBN 0-309-04161-9.

56. **The Diffusion of Biomass Energy Technologies in Developing Countries.** 1984, 120pp. Examines economic, cultural, and political factors that affect the introduction of biomass-based energy

technologies in developing countries. It includes information on the opportunities for these technologies as well as conclusions and recommendations for their application. ISBN 0-309-04253-4.

Technology Options

14. More Water for Arid Lands: Promising Technologies and Research Opportunities. 1974, 153pp. Outlines little-known but promising technologies to supply and conserve water in arid areas. ISBN 0-309-04151-1.

21. Making Aquatic Weeds Useful: Some Perspectives for Developing Countries. 1976, 175pp. Describes ways to exploit aquatic weeds for grazing, and by harvesting and processing for use as compost, animal feed, pulp, paper, and fuel. Also describes utilization for sewage and industrial wastewater. ISBN 0-309-04153-X.

34. Priorities in Biotechnology Research for International Development: Proceedings of a Workshop. 1982, 261pp. Report of a workshop organized to examine opportunities for biotechnology research in six areas: 1) vaccines, 2) animal production, 3) monoclonal antibodies, 4) energy, 5) biological nitrogen fixation, and 6) plant cell and tissue culture. ISBN 0-309-04256-9.

61. Fisheries Technologies for Developing Countries. 1987, 167pp. Identifies newer technologies in boat building, fishing gear and methods, coastal mariculture, artificial reefs and fish aggregating devices, and processing and preservation of the catch. The emphasis is on practices suitable for artisanal fisheries. ISBN 0-309-04260-7.

Plants

25. Tropical Legumes: Resources for the Future. 1979, 331pp. Describes plants of the family Leguminosae, including root crops, pulses, fruits, forages, timber and wood products, ornamentals, and others. ISBN 0-309-04154-6.

37. Winged Bean: A High Protein Crop for the Tropics. 1981 (2nd edition), 59pp. An update of BOSTID's 1975 report of this neglected tropical legume. Describes current knowledge of winged bean and its promise. ISBN 0-309-04162-7.

47. **Amaranth: Modern Prospects for an Ancient Crop.** 1983, 81pp. Before the time of Cortez, grain amaranths were staple foods of the Aztec and Inca. Today this nutritious food has a bright future. The report also discusses vegetable amaranths. ISBN 0-309-04171-6.

53. **Jojoba: New Crop for Arid Lands.** 1985, 102pp. In the last 10 years, the domestication of jojoba, a little-known North American desert shrub, has been all but completed. This report describes the plant and its promise to provide a unique vegetable oil and many likely industrial uses. ISBN 0-309-04251-8.

63. **Quality-Protein Maize.** 1988, 130pp. Identifies the promise of a nutritious new form of the planet's third largest food crop. Includes chapters on the importance of maize, malnutrition and protein quality, experiences with quality-protein maize (QPM), QPM's potential uses in feed and food, nutritional qualities, genetics, research needs, and limitations. ISBN 0-309-04262-3.

64. **Triticale: A Promising Addition to the World's Cereal Grains.** 1988, 105pp. Outlines the recent transformation of triticale, a hybrid of wheat and rye, into a food crop with much potential for many marginal lands. Includes chapters on triticale's history, nutritional quality, breeding, agronomy, food and feed uses, research needs, and limitations. ISBN 0-309-04263-1.

67. **Lost Crops of the Incas.** 1989, 415pp. The Andes is one of the seven major centers of plant domestication but the world is largely unfamiliar with its native food crops. When the Conquistadores brought the potato to Europe, they ignored the other domesticated Andean crops—fruits, legumes, tubers, and grains—that had been cultivated for centuries by the Incas. This book focuses on 30 of the "forgotten" Incan crops that show promise not only for the Andes but for warm-temperate, subtropical, and upland tropical regions in many parts of the world. ISBN 0-309-04264-X.

69. **Saline Agriculture: Salt-Tolerant Plants for Developing Countries.** 1990, 145pp. The purpose of this report is to create greater awareness of salt-tolerant plants and the the special needs they may fill in developing countries. Examples of the production of food, fodder, fuel, and other products are included. Salt-tolerant plants can use land and water unsuitable for conventional crops and can

harness saline resources that are generally neglected or considered as impediments to rather than opportunities for development. ISBN 0-309-04189-9.

Innovations in Tropical Forestry

35. **Sowing Forests from the Air.** 1981, 64pp. Describes experiences with establishing forests by sowing tree seed from aircraft. Suggests testing and development of the techniques for possible use where forest destruction now outpaces reforestation. ISBN 0-309-04257-7.

40. **Firewood Crops: Shrub and Tree Species for Energy Production.** Volume II, 1983, 92pp. Examines the selection of species of woody plants that seem suitable candidates for fuelwood plantations in developing countries. ISBN 0-309-04164-3.

41. **Mangium and Other Fast-Growing Acacias for the Humid Tropics.** 1983, 63pp. Highlights 10 acacia species that are native to the tropical rain forest of Australasia. That they could become valuable forestry resources elsewhere is suggested by the exceptional performance of *Acacia mangium* in Malaysia. ISBN 0-309-04165-1.

42. **Calliandra: A Versatile Small Tree for the Humid Tropics.** 1983, 56pp. This Latin American shrub is being widely planted by villagers and government agencies in Indonesia to provide firewood, prevent erosion, provide honey, and feed livestock. ISBN 0-309-04166-X.

43. **Casuarinas: Nitrogen-Fixing Trees for Adverse Sites.** 1983, 118pp. These robust, nitrogen-fixing, Australasian trees could become valuable resources for planting on harsh, eroding land to provide fuel and other products. Eighteen species for tropical lowlands and highlands, temperate zones, and semiarid regions are highlighted. ISBN 0-309-04167-8.

52. **Leucaena: Promising Forage and Tree Crop for the Tropics.** 1984 (2nd edition), 100pp. Describes a multipurpose tree crop of potential value for much of the humid lowland tropics. Leucaena is one of the fastest growing and most useful trees for the tropics. ISBN 0-309-04250-X.

Managing Tropical Animal Resources

32. **The Water Buffalo: New Prospects for an Underutilized Animal.** 1981, 188pp. The water buffalo is performing notably well in recent trials in such unexpected places as the United States, Australia, and Brazil. The report discusses the animal's promise, particularly emphasizing its potential for use outside Asia. ISBN 0-309-04159-7.

44. **Butterfly Farming in Papua New Guinea.** 1983, 36pp. Indigenous butterflies are being reared in Papua New Guinea villages in a formal government program that both provides a cash income in remote rural areas and contributes to the conservation of wildlife and tropical forests. ISBN 0-309-04168-6.

45. **Crocodiles as a Resource for the Tropics.** 1983, 60pp. In most parts of the tropics, crocodilian populations are being decimated, but programs in Papua New Guinea and a few other countries demonstrate that, with care, the animals can be raised for profit while protecting the wild populations. ISBN 0-309-04169-4.

46. **Little-Known Asian Animals with a Promising Economic Future.** 1983, 133pp. Describes banteng, madura, mithan, yak, kouprey, babirusa, Javan warty pig, and other obscure but possibly globally useful wild and domesticated animals that are indigenous to Asia. ISBN 0-309-04170-8.

68. **Microlivestock: Little-Known Small Animals with a Promising Economic Future.** 1989, approx. 300pp. Discusses the promise of small breeds and species of livestock for Third World villages. Identifies more than 40 species; including miniature breeds of cattle, sheep, goats, and pigs; eight types of poultry; rabbits; guinea pigs and other rodents; dwarf deer and antelope; iguanas; and bees. ISBN 0-309-04265-8.

Health

49. **Opportunities for the Control of Dracunculiasis.** 1983, 65pp. Dracunculiasis is a parasitic disease that temporarily disables many people in remote, rural areas in Africa, India, and the Middle East. Contains the findings and recommendations of distinguished

scientists who were brought together to discuss dracunculiasis as an international health problem. ISBN 0-309-04172-4.

55. **Manpower Needs and Career Opportunities in the Field Aspects of Vector Biology.** 1983, 53pp. Recommends ways to develop and train the manpower necessary to ensure that experts will be available in the future to understand the complex ecological relationships of vectors with human hosts and pathogens that cause such diseases as malaria, dengue fever, filariasis, and schistosomiasis. ISBN 0-309-04252-6.

60. **U.S. Capacity to Address Tropical Infectious Diseases.** 1987, 225pp. Addresses U.S. manpower and institutional capabilities in both the public and private sectors to address tropical infectious disease problems. ISBN 0-309-04259-3.

Resource Management

50. **Environmental Change in the West African Sahel.** 1984, 96pp. Identifies measures to help restore critical ecological processes and thereby increase sustainable production in dryland farming, irrigated agriculture, forestry and fuelwood, and animal husbandry. Provides baseline information for the formulation of environmentally sound projects. ISBN 0-309-04173-2.

51. **Agroforestry in the West African Sahel.** 1984, 86pp. Provides development planners with information regarding traditional agroforestry systems—their relevance to the modern Sahel, their design, social and institutional considerations, problems encountered in the practice of agroforestry, and criteria for the selection of appropriate plant species to be used. ISBN 0-309-04174-0.

70. **The Improvement of Tropical and Subtropical Rangelands.** 1990. 380pp. This report characterizes tropical and subtropical rangelands, describes social adaptation to these rangelands, discusses the impact of socioeconomic and political change upon the management of range resources, and explores culturally and ecologically sound approaches to rangeland rehabilitation. Selected case studies are included. ISBN-0-309-04261-5.

General

65. **Science and Technology for Development: Prospects Entering the Twenty-First Century.** 1988, 79pp. This report commemorates the twenty-fifth anniversary of the U.S. Agency for International Development. The symposium on which this report is based provided an excellent opportunity to describe and assess the contribution of science and technology to the development of Third World countries and to focus attention on what science and technology are likely to accomplish in the decade to come.

Forthcoming Books from BOSTID

Traditional Fermented Foods. (1990)

Neem. (1990)

BOSTID Publication Distributors

United States:

Agribookstore
1611 N. Kent Street
Arlington, VA 22209

agAccess
PO Box 2008
Davis, CA 95617

Europe:

I.T. Publications
103-105 Southhampton Row
London WC1B 4H, England

SKAT
Varnbuelstrasse 14
Ch-9000 St. Gallen, Switzerland

S. Toeche-Mittler
TRIOPS Department
Hindenburgstrasse 33
6100 Darmstadt, Federal Republic of Germany

T.O.O.L. Publications
Entropotdok 68a/69a
1018 AD Amsterdam, Netherlands

Asia:

Asian Institute of Technology
Library & Regional
    Documentation Center
PO Box 2754
Bangkok 10501, Thailand

National Bookstore
Sales Manager
PO Box 1934
Manila, Philippines

University of Malaya Coop. Bookshop Ltd.
University of Malaya
Main Library Building
59200 Kuala Lumpur, Malaysia

Researchco Periodicals
1865 Street No. 139
Tri Nagar
Delhi 110 035, India

China Natl. Publications Import & Export Corp.
PO Box 88F
Beijing, Peoples Republic of China

South America:

Enlace Ltda.
Carrera 6a. No. 51-21
Bogota, D.E., Colombia

For More Information

To receive more information about BOSTID reports and programs, please fill in the attached coupon and mail it to:

Board on Science and Technology for International Development
Publications and Information Services (HA-476E)
Office of International Affairs
National Research Council
2101 Constitution Avenue, N.W.
Washington, D.C. 20418 USA

Your comments about the value of these reports are also welcome.

Name

Title

Institution

Street Address

City

Country   Postal Code

Name

Title

Institution

Street Address

City

Country   Postal Code